SYSTEMS-
SENSITIVE
LEADERSHIP

SYSTEMS-
SENSITIVE
LEADERSHIP

SYSTEMS-SENSITIVE LEADERSHIP

Empowering Diversity Without Polarizing the Church

**MICHAEL C. ARMOUR, PH.D.
AND DON BROWNING**

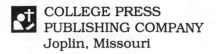
COLLEGE PRESS
PUBLISHING COMPANY
Joplin, Missouri

Library of Congress Cataloging-in-Publication Data

Armour, Michael C. (Michael Carl), 1944–
   Systems-sensitive leadership: empowering diversity without
polarizing the church / Michael C. Armour, Don Browning.
     p.   cm.
   ISBN 0-89900-736-8 (pbk.)
   1. Pastoral theology. 2. Christian leadership. 3. Values.
4. System theory. 5. Graves, Clare. 6. Maslow, Abraham H.
(Abraham Harold) I. Browning, Don, 1937– . II. Title.
BV4011.A75 1995
250'.1'1—dc20   95-2490
                                                   CIP

Contents

PREFACE

SECTION ONE: Intrapersonal Systems in the Church
1   Tension and Diversity in the Pew 1
2   The Systems Within Us 11
3   Coping with Complexity 27
4   Dominant System Transitions 41

SECTION TWO: Understanding the Systems
5   Systems 1 and 2: The Quest for Safety 51
6   System 3: The Quest for Power 63
7   System 4: The Quest for Truth 75
8   System 5: The Quest for Achievement 87
9   System 6: The Quest for Intimacy 101
10 Systems 7 and 8: The Quest for Holistic Solutions . .111
11 A Systems Summary 121

SECTION THREE: Applying Systems Insights to Your Church
12 When Systems Join Forces 133
13 The Multi-System Church 147
14 Learning to Accommodate New Systems 157
15 The Primary Systems Conflict 169
16 Understanding Resistance to Change 181
17 Diverse Expectations of the Church 195
18 Four Essentials in a Multi-System Church 205
19 Systems-Sensitive Bible Classes 219
20 Words Fitly Spoken 231
21 The Challenge of Worship 245
22 Maintaining Systems Alignment 257

SECTION FOUR: Advanced Systems Concepts
23 System Shifts and Transitions 273
24 Sustaining Systems Health 287

Preface

This book is about the "Four Big C's" of our day: change, complexity, confusion, and conflict. We are going through a period of human history when change and complexity seem to feed on one another. We change in order to deal with complexity. But change only makes things more complex. No wonder we end up confused. Nor is it surprising that conflict is on the rise. Confused people often end up at odds over the direction to take.

Where is this all headed? Where will it take us? No one knows for sure. But one thing is certain. The church is not insulated from it. The Four Big C's have installed themselves in congregational life and refuse to budge. Conflict over congregational change is on the rise, making the task of leadership both more complex and more confusing. These are indeed challenging times.

In confronting the Four Big C's, nothing has been more helpful to me than the principles you are about to study. They are derived from the lifelong research of Dr. Clare Graves, a professor at Union College in New York. I was introduced to Graves' work through Dr. Don Beck and Chris Cowan of the Human Resources Group (formerly the National Values Center) in Denton, Texas. You will find them mentioned several times in the footnotes of this volume.

Graves identified eight distinct thinking systems that shape human values and account for much of our diversity. Because they establish our outlook and priorities, these "systems within us" have an immense bearing on the "systems between us." Or as we say on many occasions, differences in our *intra*personal systems conspire to create problems in the *inter*personal systems of the church.

In the first four chapters we look at how systems diversity

leads to congregational tension. Then, in chapters five through eleven, we introduce the eight systems one by one. Chapters twelve through twenty-four are the practical section of the book. They examine congregational life thoroughly in terms of systems principles. They also outline specific strategies for preempting needless tension in the church.

The present volume grew out of a draft first developed by my co-author Don Browning. Through months of rewrite and revision, I increasingly became the wordsmith on this joint project. The final style is thus much more mine than his. But we developed every section in close collaboration, so that often the words are mine, but the insight is Don's. It has been a joy to work with him on this effort, and we both pray that our labors will bless your ministry.

The two of us owe a special expression of thanks to PenDell Pittman and Scott Johnson, our partners in Spectrum Leadership Associates, which specializes in leadership training and consulting services for churches. PenDell and Scott have an exceptional grasp of congregational systems. Their suggestions and recommendations have been invaluable in making this book a reality.

And we would be doubly remiss not to express profound gratitude to College Press for having the vision to publish a work on such a groundbreaking subject. To our knowledge, no other book has ever approached church leadership the way this one does. College Press early saw the merits of this study and wanted a part in its publication. Their words of encouragement have sustained us through these months of writing and revision, and our fellowship with them has been a true blessing.

Michael C. Armour

I always wanted to be a preacher. My Bible Class teacher ignited this desire in me when I was nine years old by standing me on the stairs leading into the baptistry of our typical white frame rural church building in our small West Texas community. She introduced me as our little preacher who is going to preach for us today. That church continued to provide me with numerous speaking occasions. By age twelve my only career focus was preaching. I longed for the day I would be a preacher.

My dream became a reality when I completed college and entered my first pulpit. I expected the unfolding years of service in the church to bring me total happiness! I was confident that my life would be characterized by unlimited joy, peace, and personal fulfillment. Somehow I had overlooked the reality that one does not do ministry in an ideal world.

It was a shock to soon discover that not everyone in the church thought and felt about things the way I did nor appreciated my sincere efforts in growing the church. As the years passed, the situation did not improve. The harder I tried, the wider the gap between my understanding and theirs became. I found it confusing and frustrating that fellow Christians did not approach most church matters as I did. I thought I could reason them into seeing things my way. Wrong! This approach increased the frequency and level of crisis. Conflict and confrontation became my constant companions. Frustrations filled my days and I began to wonder if I should continue preaching.

Then my friend and college classmate, Dr. Don Beck, invited me to a seminar he was conducting on the Gravesian concept of systems within us. It is an understatement to say I experienced an "aha!" during that seminar. The information caused lights to flash and bells to ring in my head! As I grasped the significance of systems thinking, I understood why vast differences existed between Christians. I began to understand diversity as a gift from God and a strength within the church. Most importantly, I could accept people as they were without being driven to make them think and behave like me. Conflicts and confrontations became less frequent. My newfound insights allowed peace and joy to return to my ministry again.

I wanted others to benefit from these concepts as I had. Dr.

Beck encouraged me to write a book. During the next two years I gathered material and struggled to write. It was not until I sought out my co-author Mike Armour that these concepts were framed in language which captures one's imagination. Mike has skillfully written this material in metaphors and images easily grasped by all Christians. I am deeply grateful to him for his untiring efforts with this project over the past year.

As you read, contemplate the possibilities this book raises. The least you may gain is personal understanding about why events occur as they do. In and of itself that will equip you to remain calm and composed in the presence of chaos. On the other hand, it is possible that you may discover ways for managing diversity within the church so that members can experience greater spiritual health and harmony. You may be among the first to pioneer a systems church. May God bless you as you read.

Don Browning

Tension and Diversity in the Pew

In earthquake-prone California, dozens of hospitals sit near hazardous faults. Deep in the earth deadly forces intensify daily as nature readies a furious convulsion. Above ground, meanwhile, emergency rooms prepare for that inevitable moment. Surrounding corridors bristle with life-giving technology, and trauma teams rehearse every eventuality. But a tragic irony lurks in the wings. In a highly destructive quake these very facilities may crumble in the first violent tremors. Casualties themselves, they would be powerless to help others.

As a hospital for wounded souls, the church faces a similar peril. We never know when the world's spiritual casualties will end up at our door. Ideally they will find us ready to respond with care, truth, and grace. But what if they discover the hospital in disarray, its workers so preoccupied with their own trauma that they have no energy for others?

Take Larry, for instance. Coincidence of location brought him our way. Approaching middle-age and the father of a five-year old, he had never been to church in his life. But recently, on a crisp Sunday morning, he ventured into our services. "I've begun a spiritual search," he explained, "and your church was closest to my house."

For the next hour he tried to participate, but few things made sense to him. Since he knew nothing about the Bible, most of the sermon was over his head, and the symbolism of the Lord's Supper escaped him entirely. The music was a brighter moment, for as a professional singer he could at least follow along. But the songs brimmed with biblical metaphors that for him served to confuse, not clarify.

Larry might easily have walked out, vowing never to return. Instead he promised to be back. The service was not a discouragement, for he knew he had much to learn. Besides, he had not come expecting to understand so much as he hoped to find understanding. He needed people who would sympathize with the pain of his divorce and the emptiness of his life. People who would accept his questions, even the skeptical ones. People who would never ridicule or embarrass because he knew so little.

As he slipped into the pew that morning, he was more concerned with our attitudes than our theology. He looked around for evidence of harmony and good will. He watched how people greeted one another, related to one another, found joy with one another. Had he sensed dissension, arrogance, or pretense, he would have taken his quest elsewhere.

Was Larry asking too much of us? Surely not. No one chooses a church because it is divided or in turmoil, no more than we choose a hospital because it is in ruins. We all want a church like the one Larry hoped to find — a caring fellowship where people transcend personal differences and work together in love. Every Christian pays tribute to that ideal. Larry merely expected us to take it seriously.

In the eyes of non-believers congregational strife always discredits the Christian message. Like Larry, they may have little Bible knowledge. But they instinctively sense that lovers of Christ should be lovers of one another. A young woman, herself a newcomer to Christianity, underscored this point recently. She responded in disbelief that we were writing a book to help churches avoid conflict. "Why should you need something like that for a church?" she asked.

Cannon Ball Issues

Non-believers also shake their heads at the issues we choose to fight about. All too often Christians betray their harmony for the most trivial causes, as our friend Charlie discovered in his first full-time ministry. Fresh out of college he began working with a church whose roots traced back to the early 1800s. In fact, the congregation still met in a simple stone building erected before the Civil War. The people were quite proud of their little meeting house, and everyone wanted to maintain it. Still, every discussion of repairs touched off a fracas over one contentious issue — the cannon ball.

It seems that Union and Confederate forces once fought an engagement near the church. During an exchange of cannon fire an errant round struck the building, embedding itself in an outer wall. There it remained for almost a century, half buried in the stonework. By the time Charlie arrived on the scene, many members considered this globe of rusting iron a worthless eyesore. They wanted it removed, and the sooner the better. Another group, equally vocal, would have none of that. To them this relic of war was a lasting memorial to a heroic battle, a hallowed symbol to be preserved forever. Thus, any thought of refurbishing the building was doomed from the start, all because some long-forgotten artillery unit could not shoot straight!

They Just Couldn't Agree

Years later Charlie would tell this story and laugh. "They were such good people," he would say, "but they just couldn't agree about that cannon ball." As far as he knew, they never resolved the standoff. Long after he moved away he still heard reports that the controversy was alive and well. Unfortunately, thousands of other churches have fought over differences no less inane. And in most of those cases Charlie's summary would serve as a fitting epitaph: "They were such good people. They just couldn't agree."

To be sure, agreement is not always possible, or even desirable. Some issues are simply not negotiable. If a central tenet of the gospel is at stake, we clearly cannot acquiesce. But churches rarely split over biblical absolutes. They fight about things like cannon balls.

When they do, it is easy to understand why non-believers become judgmental. After all, we look at such scenes judgmentally ourselves. We are quick to question the motives (not to mention the spiritual maturity) of people who disrupt a fellowship over mundane questions. We presume that Christians who are truly Christlike, with hearts set on crucial concerns of the faith, could circumvent conflict. Yet experience proves that assumption unfounded. Good people with good intentions can still end up at odds over non-essentials.

Take the case of two veteran missionaries, close friends and co-workers for years. Together they planted churches in a dozen cities, converting thousands in the process. But one day, as they planned a follow-up campaign, they found themselves at an impasse. The problem centered on personnel issues, namely, who else would make up the campaign team. Finally, after protracted arguments, they decided to dissolve their highly successful co-ministry and go separate ways.

How do you appraise that scenario? Another case of inadequate Christlikeness? Two people insensitive to kingdom priorities? Surely not. In case you have not noticed, we are talking about Paul and Barnabas (Acts 15:36-40). Where would we find two men more doctrinally compatible, more spiritually mature, more given to the lordship of Christ than these two? Such character and commitment, however, did not ensure perpetual harmony between them. As far as we know, the rift over John Mark was so great that they never worked together again. If even the partnership of Paul and Barnabas could succumb to conflict, we are hardly immune ourselves.

Recurring Mistakes

The last 30 years have laid our susceptibility bare. Since the early 1960s conflict and tension have escalated across all religious fellowships. Yet, churches continue to make three recurring mistakes.

- First, they often downplay their own vulnerability to conflict, rarely taking the threat seriously. Even when strife erupts in nearby congregations, leadership goes about life as usual, naively confident that "nothing like that would ever happen here."

- Second, because of that naiveté, churches pay little attention to conflict avoidance. In both long-range and short-range planning they lack a cogent method, consistently pursued, of anticipating and minimizing the risk of dissension.

- Third, when conflict does take place, churches tend to explain it in simplistic terms. Efforts to fix the problem center on a single cause (or at most a handful of causes), as though that fully accounted for the stress. A more realistic appraisal would view congregational struggles as inherently complex. Simplistic explanations do not enhance our understanding of conflict. They obscure it. In time, therefore, solutions derived from simplistic analysis tend to come undone.

Restlessness in the Pew

According to national studies, church leaders now devote 25-40% of their time to congregational tension. To their credit, they usually succeed in calming troubled waters. But not always. We could fill this book — and a dozen more its size — with accounts of churches in turmoil and leaders burned out in the process. Congregational leadership has never been easy, of course. Sin and doctrinal controversies have vexed every generation.

But contemporary churches face an added challenge. For lack of a better word, we would describe it as a general restlessness in the pew, a prevailing "dis-ease" within the body. Frequently this discontent has no immediate focus. Rather, it bides its time, waiting to latch onto just the right issue. When it finds one to its liking, it strikes opportunistically, catching leaders unaware.

As a rule this restlessness avoids brinkmanship. It may voice its complaints vigorously. But it rarely pushes things to the point of wholesale rift. Instead, it settles for a congregational equivalent of "low-intensity warfare." This is a military term for guerrilla operations that forego pitched battles. In low-intensity warfare a near-invisible enemy saps morale by mounting quick, random attacks, slipping away for weeks on end, then popping up again at some unsuspecting moment. Many leaders know firsthand the exhaustion of dealing with guerrilla-like skirmishes that break out here, then there in the church. "All I do is fight brush fires," they complain. Reduced entirely to the defensive, they see themselves as dashing from flash point to flash point, doing their best to "keep a lid on things."

When pressed to identify the source of this problem, leaders offer a surprisingly consistent answer. The primary culprit, most say, is *diversity*: too many people with too many different ideas about how things ought to be done. In one sense this is nothing new. The church has arbitrated divergent views ever since the day of Pentecost. Still, the problem is particularly acute as we approach the twenty-first century. Urbanization and technology have served up an unprecedented range of personal options, leading to highly individualized lifestyles. Diversity reigns supreme, from our choice of foods in the dairy case to our choice of jobs in the marketplace to our choice of entertainment on weekends. With personal tastes so varied, is it any wonder that we disagree on how to "do church"?

No Time for Hand-Wringing

Unless every long-range forecast is wrong, diversity will not diminish in the years ahead. It will only expand. Relentless centrifugal forces have been set loose in our society, and the church will continue to feel the strain. Faced with that reality, church leaders have but two choices. We can either wring our hands about diversity and stay on the defensive. Or we can roll up our sleeves, develop new styles of leadership, and learn to harness diversified outlooks.

This book is written for those who are ready to go beyond hand-wringing.

- The early chapters will help you understand what is driving this diversity and why it is suddenly so intense.
- Later sections then use that understanding to build new strategies for coping with diversity.

We do not propose simple or sure-fire solutions. But we do offer an intriguing way to manage congregational life. We call it "systems-sensitive leadership." This approach builds on two fruitful areas of contemporary research that together allow us to see our congregations in an entirely new light.

First is the effort to analyze human behavior in terms of systems concepts. By way of comparison, consider a parallel development in the history of science. For centuries biology looked at a pond in the woods as merely a breeding ground for thousands of species. Scientists worked to classify each of these species, then studied them individually, in isolation from one another. Although they gained vast knowledge using that method, they ultimately realized its inherent limitation. The very process of isolating an organism, they discovered, curtailed their ability to learn about it. To fully understand an organism they needed to observe its interaction with other life forms nearby. Today we no longer think of the pond as a collecting pool for myriads of isolated species. Instead, we see it as an interdependent network of living entities. We call this network an ecological system.

Congregational Systems

Once we describe a pond as a system, our focus shifts to the various components of the system and how they interact. Your focus will undergo a comparable change when you start thinking of your church in systems terms. Viewed from a systems perspective, your congregation is not merely a group of people brought together by common beliefs and aspirations. It is also a complex pattern of human networks playing off one another. In their interaction these networks define the atmosphere and dynamic of your congregation.

Networks like this never show up on organization charts. An organization chart simply shows us how to go through channels. But there is a world of difference between "going through channels" and "working the system." It is a mistake, therefore, to confuse an organization chart with the way a church actually functions. That is why the issue of *organizational structure* is secondary to systems-thinking. Of far greater interest are the human systems in your church and how they intermesh.

We are hardly the first to examine congregational life from a systems viewpoint. Several helpful and insightful books have taken a similar tack.[1] Our purpose in this volume is neither to take issue with them nor to recultivate ground they have effectively tilled. Rather, we offer a view that complements theirs. Previous studies have centered exclusively on interpersonal systems in the church. These works have drawn on systems theories that have proven invaluable in family therapy and corporate reengineering.

This study, by contrast, emphasizes intrapersonal systems (i.e., those that operate within us individually). We are dealing with skills that leaders need *before* they start applying the systems techniques other books explore. Our thesis is that *inter*personal systems flow from *intra*personal systems. To put it another way, we cannot properly understand why two people interact as they do unless we know their individual thinking and emotional systems.

Which brings us to the second area of contemporary research that undergirds systems-sensitive leadership. The past four decades have seen breathtaking discoveries about the human mind and how it works. Increasingly we have learned that we do not view things alike because we are not "wired" alike inside. Some of us are right brain dominant, some left brain. Some of us are visual learners, others aural learners. Some of us process ideas by imagining pictures, others by carrying on an internal dialogue, and still others by checking and rechecking our feelings.

In short, God did not genetically program us to think alike in every detail. Robert Fulghum says it well when he writes:

> The single most powerful statement to come out of brain research in the last twenty-five years is this: We are as different from one another on the inside of our heads as we appear to be different from one another on the outside of our heads. Look around and see the infinite variety of human heads — skin, hair, age, ethnic characteristics, size, color, and shape. And know that on the inside such differences are even greater

---

[1] The Alban Institute has been particularly interested in this approach to congregational management. Among their more recent publications are Peter L. Steinke, *How Your Church Family Works: Understanding Congregations as Emotional Systems*; R. Paul Stevens and Phil Collins, *The Equipping Pastor: A Systems Approach to Congregational Leadership*; and George Parsons and Speed B. Leas, *Understanding Your Congregation as a System: Congregational Systems Inventory*.

— what we know, how we learn, how we process information, what we remember and forget, our strategies for functioning and coping.[2]

This is not to suggest that we differ so much that we cannot agree on foundational principles, for we can. Our thinking patterns, while not identical, share considerable overlap. There is enough commonality, for instance, to permit unity on essentials of the faith (whether unity now exists or not). But there is not enough overlap to assure absolute agreement on how to implement these essentials and prioritize them. We diverge still further once we turn to more peripheral issues. And it is here, in matters outside the gospel core, that Christians often find themselves at odds.

This book will help you understand why that happens. Beyond that, it will give you new strategies for building congregational harmony and health, despite pronounced differences of viewpoint. By learning to practice systems thinking, you will also expand your problem-solving skills. Once you see the church as an intricate interaction of intrapersonal and interpersonal systems, you will avoid the mistake of explaining congregational tension in simple cause-and-effect terms. Instead, you will see the church as an extensive array of highly dynamic, interlocking processes. You will know how to delve into those processes and probe beneath the surface, where unspoken and unrecognized agendas are often at work.

Those who master the ability to view their church this way excel at foreseeing discord, then moving to disarm it. In addition they are adept at effecting change without raising the potential for conflict. Although history celebrates leaders who triumphed in crisis, the greater visionary is the one who averts crisis in the first place. Our goal is to help you be that type of leader.

---

[2] Robert Fulghum, *It Was On Fire When I Lay Down on It.* (New York: Ivy Books, 1989), p. 40.

The Systems Within Us

Systems-sensitive leadership builds on the seminal research of the late Clare Graves, a professor of developmental psychology at New York's Union College. Graves spent a lifetime studying human diversity. He was fascinated by the variety of outlooks and values that prevail across the globe. He wanted to know how these differences emerge and how they establish themselves as fixtures in the mind.

In 1974 Graves mapped his findings in an essay for *The Futurist* magazine.[1] There he described *eight thinking systems* uncovered in his research. Each system, he said, encourages a distinctive outlook on life. From one system to the next we change the way we

- define our sense of self
- organize our lives
- group our priorities
- structure relationships
- analyze ideas
- and respond to innovation and new initiatives.

Patterns of motivation also change, as do our personal approaches to leadership. Moreover, Graves discovered the following general principles.

- At birth the eight systems are latent within us.
- They activate, one by one, at various stages of existence.
- No one relies equally on all eight systems. Nor do we use them all simultaneously.

---

[1] Clare W. Graves, "Human Nature Prepares for a Momentous Leap," *The Futurist* (April 1974), pp. 72-87.

- Of the eight systems, one or two will always be so influential that they *dominate* our personal outlook.
- Moreover, any system or combination of systems can be dominant. The choice is not prescripted.
- For that reason, *dominant systems vary from person to person.*
- They also change as we move through various phases of personal development.

In a word, dominant systems are highly dynamic. They present themselves in ever-changing patterns, both in the lives of individuals and among the members of a group. The ability to recognize that dynamic and understand it is the cornerstone of systems-sensitive leadership.

Dominant Systems in the Church

Members of a church, drawn from a cross-section of the community, bring a variety of dominant systems into the congregation. Each system has its own view of how the church should go about its work, its worship, and the enterprise of fellowship. Not only that, each system has a singular style of spiritual self-expression. To downplay or ignore those differences is to set the stage for discontent and dissension. That is precisely what is happening, we believe, in churches large and small across our land. The goal of systems-sensitive leadership is to reverse this course and help churches maintain harmony, despite systems differences.

Our first step is to understand each system individually and how it approaches Christianity. Systems-sensitive leadership also requires variety in ministry style and flexibility in ministry structure. This versatility is necessary because personal motivations and priorities are not the same from one dominant system to another. And if people are not motivated alike, neither should we try to manage them alike.

Maslow and Graves

These differences in motivation, according to Graves, result from changing conditions of personal existence. As he put it, "man must solve certain hierarchically ordered existential problems which are crucial to him in his existence. The solution of his current problem frees energy in his system . . . ."[2] In other words, having resolved the needs that motivated us in one state of existence, we advance to other needs and another set of motivations.

This will sound strikingly familiar to anyone acquainted with Abraham Maslow. A contemporary and friend of Graves, Maslow showed how conditions of existence determine the types of motivations that appeal to us.[3] Maslow held that until we meet lower level needs, we do not move on to pursue higher level needs. Graves discovered that this process is far more subtle than a casual reading of Maslow might imply. New conditions of existence do not directly produce motivational change, Graves found. What they do is trigger new thinking systems, each with a unique motivational package. That is why there is an observable connection between conditions of existence and motivation.

The order in which these thinking systems unfold has some resemblance to the motivational pattern in Maslow. Thus, in the pages ahead you will see occasional points of contact between Maslow and Graves. Nevertheless, thinking systems are far more comprehensive than Maslow's scale of needs. Thinking systems account not only for what motivates us, but for interpersonal, organizational, and ethical priorities, as well. In addition, thinking systems impact how we learn, what we expect from authority figures, and how we structure institutional life. At a societal level they influence the form of government we choose and the kinds of institutions we build. Graves held that dominant systems even control biochemistry and neurological activation.[4]

------

[2] Graves, "Human Nature," p. 77.

[3] Maslow, a professor at Brandeis University, identified five fundamental categories of need that motivate mankind. The most basic of these is physiological survival. Next comes personal safety, a sense of belonging (or love), esteem (or status), and self-actualization, in that order. Abraham H. Maslow, *Motivation and Personality* (New York: Harper, 1954); also see his *Toward a Psychology of Being*, 2nd edition (New York: D. Van Nostrand Company, 1968).

[4] Graves, "Human Nature," p. 72. In my own work with multiple personality disorders, where rapid system switching occurs, I have observed impressive evidence that dominant systems do indeed affect biochemistry and neurology, as Graves said. [Mike Armour]

Application of Graves' Concepts

Many, including the American Management Association, were quick to see the value of Graves' research. As early as 1973, a year before Graves wrote for *The Futurist*, the AMA was already borrowing from him for a text on personnel management.[5] Unfortunately, Graves himself never published a truly definitive study of the systems. He died with that project unfinished. But he did describe the systems often, both in classroom and conference settings. That permitted close friends and former students to take up his mantle after his death. They have since enlarged on his initial insights and added rich detail to his basic conclusions.[6] Today counselors, educators, consultants, strategic planners, marketing specialists, and corporate executives all borrow from Graves.[7] Specialists have also used his views to great effect in highly polarized settings, from defusing gang warfare in our cities[8] to dismantling apartheid in South Africa.[9]

Now, with this volume, we are showing what Graves can mean to the life of a church. To our knowledge, this is the first wholesale effort to do so. When we first encountered Graves' concept of systems, our eyes lit up immediately. Suddenly we had a cogent explanation for the tension and diversity in our own congregations. Friction points that had never made sense became perfectly

---

[5] Charles L. Hughes and Vincent S. Flowers, *Shaping Personnel Strategies to Disparate Value Systems* (New York: American Management Association, 1973).

[6] As this book goes to press, Basil Blackwell of Oxford is preparing to publish a major volume by Dr. Don Beck and Christopher Cowan that will be the fullest and most current treatment of Graves. The tentative title for the book is *Spiral Dynamics*, based on the unique way in which Beck and Cowan depict the systems.

[7] The Federal Office of Personnel Management is drawing on Graves' concepts to help government supervisors understand their workforce. They funded the development of a text by Christopher Cowan entitled *Human Dynamics — Managing Differences through an Understanding of Value Systems* (Denton, TX: The National Values Center, 1989). Also *Implementing Self-Managed Work Teams* (U.S. Office of Peronnel Management, Dallas Region, 1994).

[8] For example, Roger Ruth has used Graves' systems for years in dealing with inner-city youth problems in Tulsa, Oklahoma.

[9] Don Beck and Graham Linscott, *The Crucible: Forging South Africa's Future* (Denton, TX: New Paradigm Press, 1991).

understandable. Not only that, we now had a way to anticipate that friction and minimize its impact. We have written *Systems-Sensitive Leadership* in the hope of sharing that thrill of discovery with you.

But be forewarned. You will be halfway through this book before the tone turns practical and you find specific applications for your church. Prior to that point we will be explaining the systems themselves. This may seem an excessively long introduction, but experience has taught us not to rush it. Proficient systems skills require a considerable knowledge base, and we want to lay that foundation carefully. As we said in chapter one, we are not advocating a simple solution to church conflict. But we are proposing one rich in rewards for those willing to master it.

Conceptual Systems and Cohesiveness

So where to begin? Perhaps we should start by making a clarification. When we speak of dominant systems spawning diversity in the church, we are not talking about doctrinal diversity, at least not in the strict sense of the word "doctrine." Dominant systems do not dictate the *content* of our thinking (i.e., *what* we think), but they do shape our thinking *style* (*how* we think).

For instance, some systems are highly kinesthetic, tapping into feeling and emotion. Others elevate logic and analysis. Some lean toward self-interest and independence, while others foster identification with a group. In effect we "conceptualize" differently from one thinking system to another. Throughout this book, therefore, we will speak of "conceptual systems" and "thinking systems" interchangeably.[10]

So long as members of any group have the same dominant system, they tend to approach issues alike. This does not mean they see eye-to-eye on every specific. They may not. But in terms of styles, methods, and structures they generally go at things alike

---

[10] Graves himself preferred the term "value systems," for we use them to "evaluate" experience. To most people, however, the word "value system" connotes personal ethics and morals, not at all what Graves had in mind. To reduce the risk of confusion, we have opted to forego Graves' naming convention.

because they look at things alike. This gives the group a natural cohesiveness.

Once dominant systems diversify, cohesiveness starts to un-ravel. People may still agree on the "facts" (or in the case of a church, on doctrine) but be worlds apart on what those facts mean. Disagreements over methodology become more frequent and more pronounced. Differences in personal preference become an increasing distraction. People are drifting apart conceptually, not so much on basic beliefs, but on how to *package* those beliefs. You see this in congregations where members agree on core doctrine and commitments, yet are at odds over how to organize the life of the church, prioritize its efforts, structure its Bible classes, and conduct public worship.

Similar problems occur, of course, when you mix people with different personality types. We must caution, however, against viewing these thinking systems as a way of typing people. They are not. We will elaborate on why they are not in chapter twelve. For the moment let us merely say that systems differences are far more profound than those that stem from personality types, as you soon will see.

Corporate Culture

The ability to maintain cohesiveness in a church depends largely on its "corporate culture." Once enough people in a group share a common dominant system, the outlooks and values of that system start to characterize the group itself. Or to put it another way, corporate culture always reflects the distinctive influence of the prevailing dominant system among the group's members. This is true whether we are speaking of institutions, organizations, congregations, or even entire civilizations.

Thus, your church leans toward some dominant system, as does each of its Bible classes, all of its ministry teams, and every committee in the congregation. Not everyone in the group may have the same dominant system, to be sure. But the group will have a distinct system overtone, nonetheless. The people we call "misfits" are often those whose dominant system simply does not coincide with the primary system of their group. If the difference is not too pronounced, we may label them merely as "odd" or "a bit strange." At more significant levels of difference we resort to words like "maverick," "troublemaker," or "dissident." Ultimately the names "traitor," "turncoat," and "heretic" come into play.

Systems-sensitive leaders are careful to distinguish people who are truly trouble-makers from those whose dominant system sets them at odds with the rest of the group. The role of leadership is to foster a corporate culture that not only recognizes the value of the various systems, but gives them an opportunity to express themselves constructively. What commonly happens is that the corporate culture is too inflexible to embrace the minority systems in its midst. Those non-majority systems then must either acquiesce (which leaves them disgruntled) or else depart.

Why All These Systems

The reason there is such discontent in the church today is that the number of systems in the pew has suddenly burgeoned. Thirty years ago, had we surveyed most congregations, we would have found only one or two dominant systems among its members. (This will become self-evident to our older readers when we come to specific system descriptions later.) Now, as our own evaluative work with congregations has shown, we are likely to discover three, four, or even five. By the end of this chapter you will understand why that has happened. The net result is this. While the church has always contended with diversity and tension, those forces are now at unprecedented levels. Congregational leadership must reconcile a range of views and outlooks unlike any we have seen before.

But why do we need all these systems in the first place? What purpose do they serve? In principle, they are elaborate coping mechanisms. They help us deal with the complexity of our personal existence. They give the mind the elasticity it needs to make sense of life, no matter how complicated our world becomes.

Early in life only one or two systems are active. Conceptually these systems use relatively simple categories to sort out our existence. Thought at this level contains little ambiguity and turns to abstraction quite sparingly. Yet, despite this simplicity, such modes of thinking are all we need for our earliest states of existence.

There comes a point, however, when complexity overwhelms these initial systems. They no longer provide a conceptual framework that is adequate for daily existence. Recognizing this danger, the mind triggers more comprehensive thinking systems to help us sort out experience. Like the immune system, which creates an antibody only when some invasive threat makes it necessary, our mind does not energize more complex thinking systems until it has a need for them.

Figure 2-1 depicts the result of this process. The line at the bottom shows the complexity of existence increasing from left to right. At lower levels of complexity we draw on conceptual systems with the qualities listed on the left. As existence becomes more complex, we turn to systems that show more of the qualities on the right.

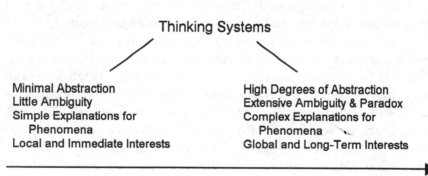

Minimal Abstraction
Little Ambiguity
Simple Explanations for
   Phenomena
Local and Immediate Interests

High Degrees of Abstraction
Extensive Ambiguity & Paradox
Complex Explanations for
   Phenomena
Global and Long-Term Interests

COMPLEXITY

Figure 2-1: Thinking System Contrasts

This chart could be misleading, however. It might suggest that more comprehensive systems — those diagrammed to the right — *replace* the ones on the left. In reality, something more subtle occurs. As new systems activate, they build cumulatively on those already present. This offers distinctive benefits, as we shall see later. For one thing it permits wholesale flexibility. When we engage in simple activities (like sports or games), we can shift toward the left side of Figure 2-1. When we turn to something more complex (like developing an international marketing plan), we can make a move toward the right.

The Complexity of Personal Existence

This freedom to draw on multiple systems does not override the fact that only one or two will prove dominant. Later we will see how culture and upbringing have a voice in our choice of dominant systems. Graves believed that nature and genetic encoding also played a role.[11] But the greatest determinant of dominant systems is the complexity of our personal existence.

And we must underscore the phrase *personal existence*. In settling on a dominant system, our mind takes its cues, not from complexity *per se*, but from our own *perception* of complexity. To put this another way, in choosing our dominant system, the mind responds to the complexity of our *existence*, not the complexity of our *world*.

- The complexity of the *world* is an *objective* reality, outside of me.

- The complexity of my *existence* is a *subjective* reality. It is a product of what I have experienced, sensed, and felt.

Living in an equally complex world is not the same as facing an equally complex existence.

To illustrate, consider the following pair of stories. Neither of them involves a change in dominant thinking systems. But they demonstrate how we respond to the complexity we perceive, not the complexity that prevails. Recently a detective on a big-city police force recounted his frustration with a new assistant district attorney. She kept throwing out cases, telling his officers, "People don't do the kinds of things you wrote in this report."

---

[11] Clinical observation makes it clear that mental impairment can prevent the more comprehensive systems from activating. Neuro-chemical imbalances can have a similar effect. As we counsel people with obsessive-compulsive disorders, we see them endlessly repeat rituals to soothe anxiety. This type of behavior is characteristic of what we will refer to later as System 2 thinking. To cope with their level of complexity, these people need to function in System 4 or 5. But without pharmacological intervention (which changes the brain's neuro-chemistry), many of them seem trapped in System 2, unable to move on.

After that happened a few times, someone looked into her background. The daughter of a wealthy New England family, she had attended elite private schools all her life. From a small, prestigious women's college she went directly into an Ivy League law school and from there to her assistant D.A.'s post. In essence she had never encountered the world the police were battling nightly.

Their solution was to persuade her to make evening patrols with them. "Maybe you can show us how to file our reports more appropriately," they explained. Their real intent, as you can imagine, was to treat her to a dose of reality. They took her into the toughest part of town, responding to one call after another. Before long her perceptions shifted significantly. She began to give more credibility to the charges that crossed her desk. In the course of those patrols the complexity of her world did not change. What changed was the complexity she perceived.

Then there was the case of a young man in our church who had a simplistic explanation for the homeless problem. People who lived on the streets, he argued, were nothing but ne'er-do-wells, too lazy to work or learn. One night some friends talked him into joining them at a downtown soup kitchen, where they worked as volunteers. Later, as they started for home, the young man fell strangely silent. When asked if something was wrong, he confessed to being troubled. "You know the fellow who came through the line in a blue knit cap?" he asked. "He was my suite mate in college. The last time I saw him, he was knocking down a six-figure salary." Suddenly the homeless problem was not so simple to explain.

Perceptions of Complexity

Thus, at a given moment complexity is a constant across any organization, society, or culture. But the *perception* of complexity will vary widely from person to person. A number of tributaries flow into our perception of complexity. Some of the more common ones are:
- the proximity of dissimilar people
- natural, economic, or political disaster
- personal health crises

- personal or family trauma
- education
- expanded knowledge of the world, its cultures, and its problems
- new technology

Even something like newfound stability or prosperity can add to our sense of complexity by giving us leisure time to reflect on issues that we previously pushed aside. When we begin to study those issues, we often uncover complexities we never realized before.

Because perceptions of complexity are not uniform across all human societies (or from person to person within a society), our internal systems trigger in highly individualized patterns. Four members of a church may live in the same neighborhood, but function from four different dominant systems. Like that assistant district attorney and the detectives, they may be part of the same complex world, but not sense the same degree of complexity.

Complexity and Diversity

Generally speaking the tie-in between perceived complexity and activated systems explains why diversity has become so pronounced of late. Until fairly recent times only a few visionaries and pioneering spirits recognized the need for several of the systems that are today commonplace. The average person had a simple view of nature, causation, and disease. Social structures were likewise simple, and knowledge of other cultures was limited. In a world like that, people could manage their existence without calling on the most complex systems. With only a handful of systems active, diversity was somewhat restricted, both intellectually and socially.

For the typical worker, those conditions of existence continued well into this century. We are less than three generations removed from a time when only three or four systems governed the thinking of the American workforce. But our existence has rapidly become so complex that four more systems have burst on the scene. For the most part this has occurred in the general populace since the Second World War, most dramatically in the last 30 years. Now people choose their dominant system from among six or seven contenders, not three or four.

Put simply, the last half-century has carried us over a systems-triggering threshold unlike any we have crossed before. Historically a system became dominant in a culture over a protracted period spanning generations. Contrast that to our own day, when several new systems have become ascendant overnight. Their meteoric appearance signals the mind's effort to cope with the equally sudden explosion of technology, urbanization, and split-second global communication — in a word, complexity.

In one of our seminars recently Mike Armour reminisced about his father. Here was a man, born in the Indian Territory, whose family moved to Texas in a covered wagon. Yet, before he was seventy years old, he watched men walk on the moon. No previous generation could have imagined such change, much less have thought it possible in a single lifetime. Still, it happened, and the mind has had to accommodate it. Little wonder that new conceptual systems, capable of managing greater complexity and ambiguity, are rapidly gaining sway. Nor is it a wonder that these newer systems are also showing up in the pew.

Reinventing the Rules

This admixture of dominant systems is altogether unprecedented. Throughout most human history there has been a certain homogeneity in adult thinking patterns across a society. But no more. There seems to be a threshold of complexity (and we have clearly crossed it in the last few decades) when four or five dominant systems must learn to coexist in the same social setting.

Unfortunately, we have no experience in managing this type of diversity. We are having to work out new rules of leadership. Because we are so early in that learning curve, institutions of every stripe, including churches, are feeling increased tension and polarization. That is why systems-sensitive leadership has become urgent. To avoid disarray, churches must harmonize emerging dominant systems with those that have long existed. Leaders must learn to maintain healthy partnerships among these competing systems, recognizing that each system has unique strengths and drawing on those strengths appropriately.

Of course, we can always deal with systems tension by defining ourselves narrowly, then letting disgruntled elements go their own way. Many organizations have found that approach too enticing to resist. The deep polarization in contemporary society and politics is one product of that strategy. But such options are not open to the people of God. Unlike other institutions, the church has a biblical mandate to maintain unity. We have no choice but to preserve cohesiveness despite diversity.

Encoded for Diversity

How, then, should we proceed? First, we must forego any thought of having everyone use the same dominant system. Neither enticement nor coercion can bring that about. People change dominant systems only when issues of existence overtax their previous system. As leaders we have no power to bring that about. We may create opportunities, like the police did with the assistant D.A., for people to change their perceptions of complexity. But if they do not, we cannot force the change.

Second, we must see God's hand in what we are experiencing. Since nature placed these systems within us, God is apparently their author. By His design the mind has an inherent ability to restructure itself in order to cope with a complicated existence. This means that diversity is a gift from God. We must therefore see diversity as a blessing, not a curse. And as Paul warned the Corinthians, we must never let His gifts become sources of discord.

Third, if diversity is God-given, the tension it yields is also from Him. Admittedly this is an unusual way to look at things. We tend to think of tension as counterproductive, a negative in congregational life. But tension as such is neither good nor bad. What we do with it determines its merits. Every field of creative endeavor uses tension to advantage. To offer a simple engineering example, we could not build magnificent archways without the tension of the keystone. Nor could we bridge a chasm without structural tension in the spans. Similarly, a congregation with no creative tension lacks the spark for vision, imagination, and fresh insight. Such a church easily becomes lifeless, no longer a body, but a corpse.

Diversity As Potential

When congregations fail to manage tension properly, they easily end up at war with themselves. Parties draw swords, then slip into an "us-versus-them" mentality. As a rule both sides attribute the conflict to differences in spirituality, maturity, knowledge, or reasonableness. (Of course, each party considers itself the paragon of these virtues.) Overlooked is the possibility that we differ simply because our minds do not work alike. And the reason they do not work alike is that God designed us that way. It seems safe to presume, therefore, that He intended divergent viewpoints to be constructive, not destructive, a source of strength, not division.

The key is to harness tension creatively. The role of leadership is not to eliminate diversity, but to capitalize on it, to harmonize it like a composer scoring a symphony. The word "harmony" presupposes an underlying divergence. An orchestra becomes harmonious by blending distinctively different notes. Without those differences the instruments can play in unison, but they cannot fashion harmony. While we prefer unison to sheer noise, what truly stirs the soul is an exhilarating interplay of beautifully developed chords.

Diversity and God's Purpose

This brings us, then, to the true strength of systems-sensitive leaders. They know how to use diversity for the purpose God intended. Due to their systems perspective, they have a unique understanding of why we are diverse and how that affects our outlooks. They also know that each outlook offers something special to the work of the kingdom. Systems-sensitive leadership seizes on that "something special" and gives it a place for cooperative, creative expression. When that happens, the church positions herself for new heights of effectiveness. Not only that, she moves closer to what God called her to be when He endowed her with diversity from the start.

We must keep in mind that ethnic differences in the early church were as troublesome as systems differences today. Yet Scripture challenged the church to transcend the clash of views between Greek and Jewish outlooks. The central thesis of this book is that we must do the same with systems diversity in the modern church. God wants diversity to be our ally, not our adversary, a force that empowers, not one that polarizes.

3

Coping With Complexity

The mind is a factory that designs and refurbishes mental models. It uses these models to make sense of what we experience. Over a lifetime we generate thousands of them. Some define our sense of self. Some help us navigate busy intersections. Others help us plan a career. We have models for every facet of life. In fact, we only know what something means once we have a model to account for it.

That being the case, the complexity of our existence must never outdistance the complexity of our models. Otherwise life will quit making sense. On the other hand we have no use for highly involved models if our circumstances are simple and uncomplicated. In that situation intricate models serve only to confuse. What we need, therefore, is flexibility, the freedom to choose models suited to our moment in life.

That is precisely the function of the eight thinking systems. Each system is a specialist, fashioning models of unique style and complexity. We can arrange these systems, as we have in Figure 3-1, according to the complexity of their models. In this sequence System 1 constructs the most basic models, System 8 the most intricate ones. In between, models grow progressively more complex each step of the way. By the time we get to Systems 7 and 8, models are so elaborate that few people currently need them.[1]

---

[1] We have chosen to identify the systems by number. Graves used a two-letter designation for each system. Don Beck and Christopher Cowan have popularized a color scheme that gives a specific hue to every system. A few people have tried to give actual names to the systems, but this invariably comes up short. The systems simply embrace too many realities for a single word to describe them.

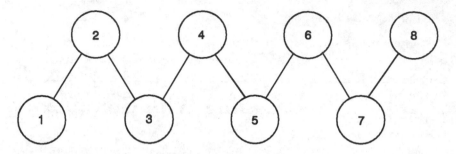

Figure 3-1: The Thinking Systems

And that is a key phrase — "need them." As we saw in chapter two, these systems lie dormant within us at birth. Our minds activate them one by one as our personal existence grows more complex. We may go through protracted periods, perhaps stretching over decades, when no new systems become prominent. This indicates that our current models continue to explain the world to our satisfaction. So long as that is the case, we have only limited interest in models of greater complexity. Intellectually we may understand them. Theoretically we may acknowledge their merits. But existentially, they will serve no purpose for us.

Odd-Even Contrasts

You will notice in Figure 3-1 that the odd-numbered systems align themselves along the bottom of the graphic, the even-numbered ones along the top. We have arranged them this way to emphasize an important peculiarity. Odd-numbered systems mark phases of our existence when individualism, self-expression, and self-reliance are driving forces. During those periods we basically derive our identity from personal accomplishment and success.

Even-numbered systems are characteristically times when we literally "regroup." At these stages of life we define ourselves in terms of a group whose identity or cause becomes our own. The even-numbered systems also involve intellectual and emotional regrouping, for they represent times of introspection, reflection, and a rethinking of critical issues.

Developmental Rhythms

As we move sequentially through the systems, each new system allows us to build a host of new models. But because odd-numbered systems are so given to self-expression and individual initiative, we tend to diversify models more rapidly (along with personal choices in behavior) when odd-numbered systems are dominant.

Not only that, the expansion of personal options in an odd-numbered system causes old social rules to break down. The way our group has done things in the past (whether that group be our tribe, our family, our nation, or our church) no longer seems adequate. This is the *Fiddler on the Roof* syndrome, where former ways make no provision for things that now beckon. Eventually, therefore, the expressionism and individualism of an odd-numbered system require us to rethink the rules that bind our community together.

To work through that issue, we move toward the next even-numbered system, where another period of "regrouping" occurs. There we redefine what we need and expect from community-building relationships with others. Because of this "group" focus, even-numbered systems are generally conservative in nature. They call for submission to the overall purpose and mores of the group. Individualism, if not discouraged, is greatly subdued.

In terms of their impact on social units, therefore, we can think of Systems 3, 5, and 7 as *change* systems, Systems 2, 4, 6, and 8 as *coalescing* systems. (System 1 has no real social component, as we shall see when we describe all these systems later in the chapter.) Our personal development thus follows an alternating rhythm in which we move from dominant systems that are self-expressive to those that are group-oriented and back again.

There are exceptions to this rule, however. People with strong egos and high needs for self-expression may skirt the "group systems" and track through the odd numbers most of their life. Once System 3 becomes dominant, they go to System 5 for their next dominant system, with only a passing glance at System 4. From System 5 they may eventually move to System 7 without stopping at System 6. On the other side of the ledger, some people track primarily through systems with even numbers. These are usually individuals who are quite cautious by nature and prefer not to venture too far on their own. The conservative tone of the even-numbered systems has an intrinsic appeal to them.

System Activation

In advanced societies like the United States, the first four systems are normally vibrant by early adolescence. System 1 begins to function the moment we are born. System 2 starts its rise to dominance about the time children start putting words together in sentences. System 3 follows suit in the early elementary years. Then System 4 begins to assert itself about the time of puberty. The remaining four systems normally do not become dominant until some time later, if at all. It is generally more difficult to anticipate when Systems 5 through 8 will energize than it is for Systems 1 through 4.

Americans born in the 1940s illustrate this point vividly. By the time they reached midlife, Systems 6 and 7 were dominant for millions of them. But that pattern was hardly universal. Much of that generation entered the middle years with little or no evidence of System 6 or System 7 influence. *The absence of those systems had nothing to do with intelligence or maturity.* These same individuals were often highly successful, psychologically stable, and emotionally mature. But their personal existence did not require the more complex systems.

Core Models

We should also point out that mere activation of a system does not mean it will ever be prominent. A particular conceptual system can exist for decades as nothing more than a marginal factor in values and outlooks. For example, Systems 5 and 6 may illuminate in an individual, and even come to exert frequent influence on decisions, yet never supplant System 4 as the dominant system.

To understand why this happens, we need to distinguish between two types of models. We might call them "casual models" and "core models," based on their relationship to our core being.

Casual models are by far the larger group. But they also have the least influence on our sense of self and personal identity. We revise and discard these models routinely with nothing more than "casual" impact on our existence. Even when a casual model has been deeply entrenched, we may replace it instantly with another on the basis of nothing more than new information.

Not so with core models. They are far more persistent, and rightly so. Their impact on us is anything but casual. These are "essential models" in the sense that they define the very essence of who we are. They address such questions as:

- How should I assess my self-worth?
- What is my purpose for being here?
- What is my role in society?
- What principles should guide my most crucial decisions?
- What do I need from others around me?

We are understandably guarded when it comes to revising these models. They are vital to psychic survival. We therefore change them at a measured pace and only in incremental steps.

As a consequence, when new conceptual systems first illuminate, we use them more readily for casual models than for core models. Not only does this ensure psychological stability, it also adds to our flexibility. We can benefit from a new thinking system without having to embrace it *in toto*. This permits us to draw on *all* activated systems for *intellectual* questions, while relying on dominant systems for *existential* questions.

Thus, what causes a change in dominant systems is not simple overtaxing of models. Overtaxed *core* models are the key. So long as our core models seem adequate for our most pressing existence issues, we see no need to consider another dominant system. We view ourselves as doing quite well right where we are.

What unsettles this state of affairs is the weight of new threats to our core existence. When we say these threats are new, we do not mean they have never existed before. Rather, they have never been an urgent concern. They are new from the standpoint that we are highly sensitive to them for the first time. If our core models cannot cope with this new sense of danger, we start looking for others that can.

Which brings us full circle to something we said in chapter

two. There we described thinking systems as coping mechanisms. The word "coping" implies a threat to safety or well-being. Every system, it turns out, is structured to contend with a cluster of threats, all grouped around a single concern. We call this the system's "primary existence issue." This issue may be physical survival. It may be making peace with the unseen forces behind the universe. It may be fending off my warlike neighbors. But every system has one such concern.

Within each system the primary existence issue determines how we define personal security and significance *within that system*. Because no two systems focus on the same primary existence issue, neither do they end up with identical definitions of security and significance. That is why there is such a broad-based conceptual shift from one thinking system to another. No one can redefine what it means to feel secure and significant without a wholesale effect on core models.

A Systems Overview

Perhaps the best way to explain this is to describe each system individually, identifying its primary existence, security, and significance issues. We will offer a more detailed introduction to these systems beginning in chapter five. Therefore, do not distract yourself at this point by trying to remember which systems and issues go together. You will pick that up naturally as we proceed. For now merely notice the general contrast between systems. That will help you see why our worldview shifts significantly as we move from system to system.

System 1

Primary Existence Issue: Physical survival in the face of immediate threats to my very life.

Basic Life Assumption: Nothing else matters if I do not stay alive.

Fundamental Security Issue: I feel secure if I have the minimal necessities of food, water, and warmth.

Fundamental Significance Issue: I am significant if I can find nourishment and protection from the elements *today*.

System 1 provides the basic physical instincts for survival. Active at the moment of birth, it alone governs our first few months. In System 1 we do not *live* life so much as we *react* to it. Our entire outlook is wrapped up in satisfying basic appetites — hunger, thirst, the need for warmth, etc. Although we move beyond System 1 in early childhood, we may return to it in extreme life-threatening conditions or in the final stages of terminal disease.

System 2

Primary Existence Issue: Personal safety in a world of unseen powers.

Basic Life Assumption: Standing behind the events of nature are spiritual forces (normally hidden from sight) that are vastly more powerful than I am.

Fundamental Security Issue: I feel secure if I can make peace with the unseen powers that control the world.

Fundamental Significance Issue: I am significant if I can achieve a blessing from the unseen powers, or at best, avoid their curse.

When System 2 appears, safety replaces survival as our driving impulse. For System 2 the primary quest is to determine if the world (both seen and unseen) is trustworthy and can be counted on for physical security. In working out that issue, System 2 employs a rich imagination. It tries to envision the invisible powers behind the forces of nature. For the child this takes the form of fairy tales, in which good triumphs over evil. For ancient cultures, mythology was a similar enactment. The world is a "magical" place for System 2, filled with wonder, amazement, and spiritual potency. But System 2 always feels vulnerable, at the mercy of things unseen. It therefore likes predictability, routine, and ritual. These assure System 2 that things are the same today as they were yesterday, which means all is well. System 2 finds a particular sense of security in group ceremonies and rituals. Thus, where System 2 prevails, tribe-like organizations emerge, with minimal hierarchy, a sharing of "sacred ground," and little sense of private ownership.

System 3

Primary Existence Issue: Physical safety in the face of hostile human forces.

Basic Life Assumption: In a dog-eat-dog world I better watch out for number one. Otherwise I won't be around very long.

Fundamental Security Issue: I feel secure if I can overpower and control anyone who poses a threat to me.

Fundamental Significance Issue: I am significant if I am tough, able to stand on my own two feet and face danger without flinching.

System 3 looks for security in strength. System 3 is a response to life conditions in which other human beings pose a serious physical threat. In those kinds of circumstances, System 3 believes, you are either a winner or a loser. And System 3 clearly intends to win. System 3 remains strong at the expense of others. It is quick to take advantage of weakness and it exploits vulnerability. In a world governed by System 3 the fittest survive and the rest perish. Not surprisingly, this is the primary system we use to fight wars, build empires, and police prisons. Much like a military structure, System 3 organizations take the form of a pyramid, with powerful decision-makers at the top and everyone else arranged in a precise pecking order.

System 4

Primary Existence Issue: Moral and social stability in a world given to hedonism, impulse, passion, and violence.

Basic Life Assumption: There is a realm of Truth, replete with enduring principles of what is right and what is wrong. My steadfast commitment to Truth and its cause assures that goodness will ultimately triumph in the earth.

Fundamental Security Issue: I feel secure when I discover timeless principles, then align my life with them.

Fundamental Significance Issue: I am significant if I remain faithful to Truth and to what it demands of me.

As opposed to the individualism and self-expression of System 3,

System 4 stresses duty, loyalty, responsibility, and accountability. System 4 believes society itself to be at risk if System 3 impulses remain unchecked. As a consequence, System 4 stresses self-discipline and postponed gratification. It defines its priorities around truth, justice, and fair play. System 4 tirelessly explores values and beliefs, the "why" behind the rules.

Because it sees the threat from System 3 as highly pronounced, System 4 looks for strength in numbers. It organizes around tightly-knit groups of "true believers" who band together against a hostile world. System 4 is quick to point out infractions, using guilt and fear to keep behavior in line. On the other hand, it is also highly self-sacrificial, the first system in which we become impassioned for causes, even to the point of dying for them. (In previous systems we have passion for people or things, but not causes.) System 4 organizations thrive on personal dependability and commitment. They make clear distinctions between right and wrong, with policies and procedures for everything. They are also guarded about change, fearful that System 3 will exploit times of transition to regain an upper hand.

System 5

Primary Existence Issue: Personal success and achievement in a world whose demands for conformity thwart my inner sense of fulfillment.

Basic Life Assumption: To develop my abilities and potential, I need a place where I am free to innovate, try new things, and explore endless possibilities.

Fundamental Security Issue: I feel secure if I can outdistance others by being well-informed, quick on my feet, timely in decisions, efficient in my efforts, and tireless as an innovator.

Fundamental Significance Issue: I am significant if I have attained impressive goals and have surrounded myself with things that bespeak my success.

System 5 emerges as a counter-balance to System 4's caution and control. System 4, with its love for stability and steadiness, is not particularly friendly to innovative spirits. They suffocate under it.

System 4 can also crush those who are overwhelmed with feelings of failure and cannot bear up under a steady message of guilt and fear. System 5 is thus a response, not to threats of physical survival (as in Systems 1, 2, and 3) or threats to moral and social survival (in System 4), but to psychological survival.

System 5 reacts to System 4's constraints by pursuing personal potential. System 5 is an avid resume builder. It packs life full of activity and achievement. System 5 idolizes effectiveness and efficiency, and it wants freedom to prove its capabilities on every turn. This is the driving system behind the entrepreneur. It is also the system that gave rise to modern science and technology. With its strong bias toward achievement and effectiveness, System 5 spawns organizations that emphasize strategic planning, goal-centeredness, and well-managed time, along with major rewards (usually material in nature) to those who perform well.

System 6:

Primary Existence Issue: Building bonds of intimacy and mutual support in a world given to insensitivity, alienation, and exploitation.

Basic Life Assumption: Only a life shared intimately *with* others and *in the service* of others has ultimate meaning.

Fundamental Security Issue: I feel secure if I can be part of a group that cares deeply for me, as I do for them, whether we are related by ties of blood and ethnicity or not.

Fundamental Significance Issue: I am significant if I care deeply enough to shoulder the load for those whom tragedy and trauma have crushed.

System 6 energizes when we awaken to the price the world has paid for System 5 achievement. In its rush to build a resume and a profitable bottom line, System 5 can run roughshod over people, relationships, and the environment. System 6 tries to bring healing to that injured world. The ecology movement comes out of System 6, as do recovery groups, the various "rights" movements, and the press for total egalitarianism. System 6 seeks bonding in place of alienation, intimacy in place of effectiveness, and compassion in place of profits. System 6 organizations typically function in small cells, often consisting of no more than 20 people, with decision-making dispersed throughout the team. In its spirit of egalitarianism, System 6 minimizes both hierarchy and the distinctions between managers and those they manage.

System 7

Primary Existence Issue: Averting the looming disaster of a polarized world in which rigid viewpoints and partisan spirits promote warring camps and thwart the flexibility we need to survive.

Basic Life Assumption: In a world permeated by uncertainty, we must be endlessly adaptable, not locked into a single way of doing things and forcing people to conform.

Fundamental Security Issue: I feel secure if I maintain a holistic perspective, having enough insight, foresight, and flexibility to anticipate change, stay ahead of it, and live harmoniously with others in the process.

Fundamental Significance Issue: I am significant if I see beyond surface confusion to underlying patterns and realities, then work to align forces that otherwise labor at cross-purposes.

In contrast to System 4, which seeks stability and order, System 7 presumes that change is a constant and cannot be avoided. System 7 therefore tries to "go with the flow" of reality. That is, it wants to pass lightly through the earth. It strives to understand physical, spiritual, and social processes, then work in harmony with them. System 7 is very "situational" in its approach to strategy and organization. If circumstances suggest a rigid hierarchy is appropriate, that is what System 7 promotes. Tomorrow, however, as circumstances change, it may push for a very egalitarian structure, the kind we would find in System 6. This flexibility at the heart of System 7 leads to frequent, but intentional midcourse corrections. To know what corrections to make and when to make them, System 7 thrives on masses of information. It is a data hungry system. It loves trend analysis. It wants to look over the horizon to see what is coming and be ready before it arrives.

System 8

Primary Existence Issue: Creating a genuine sense that all humanity is one race, living in a single village, providing equal access to the planet's resources, but caring for the earth as a fragile life-partner.

Basic Life Assumption: To assure continued survival we must create a sense of "oneness" with all living things, recognizing the rich mystery behind life itself and working to preserve life in all its forms.

Fundamental Security Issue: I am secure if I can undo the destructive rivalries and parochial short-sightedness that have pitted men against men and mankind against nature herself.

Fundamental Significance Issue: I am significant if I genuinely identify with all human beings as my brothers and sisters, regardless of ethnic or nationalistic identities, and work with them to sustain the life-giving well-spring of the planet we share.

System 8 has come on the scene so recently that we still have much to learn about it. We do know that it views the world as a single global unit. It looks at the planet the way a camera does from outer space, i.e., without seeing national boundaries and ethnic enclaves. System 8 does not deny that such things exist. It simply has greater loyalty to the entire human race than to any nation or ethnic group. For humanity to survive, System 8 believes, we must recognize how precarious life itself is. Relentless exploitation of the eco-sphere cannot continue without destroying us all. System 8 thus works to create global alliances aimed at harmonizing man and nature. It also recoils from the notion that existence is merely a material phenomenon. It perceives a universe in which a non-material reality permeates both the infinite and the infinitesimal. System 8 approaches that reality with a respect that recaptures the awe and mystery of System 2.

Complementary Viewpoints

As we stepped through these system descriptions, perhaps you noticed that the primary existence issues moved from survival of the individual to the survival of humanity, from "local" and immediate concerns to global and long-term concerns. Outlooks changed from the rather simple "us-versus-them" mindset of Systems 3 and 4 to the commitment to inclusivism in System 6 and planetary harmony in System 8. The emphasis on timeless ritual in System 2 and timeless truth in System 4 contrasts with the world of System 7, where few things are nailed down and change is constant.

Now, none of these are contradictory ways of viewing reality. They are complementary ways of doing so. There are times we need each of them. If ruffians accost us on a street corner, System 3 is most helpful. At a moment like that we can make good use of a simple, black-and-white, win-lose, me-versus-them view of the world (not to mention the surge of adrenaline it gives us). For analyzing all the variables in a global economy, however, System 3 is of little use. There we might find System 7's comfort level with ambiguity and uncertainty far more appropriate.

What you can see, however, is that with each succeeding system there are far more tension points to resolve in the primary existence issue. That is why these systems create models that are progressively more complex from System 1 to System 8. We can describe the complexity of a model as the degree to which it

- deals with larger scale problems
- involves ambiguity and paradox
- brings diverse elements together
- introduces large numbers of variables
- incorporates long-range considerations
- and requires vast amounts of information to maintain

The later systems are thus not a measure of greater intelligence, fuller maturity, or moral superiority. They are merely a more complex way to look at the world and the realities it unfolds.

Before we get away from this overview we should also note that each system has more than a single existence issue. It entails dozens of them. But one way or another, they are all fundamentally an extension of the primary existence issue. It represents the one overarching concern that brings a system to dominance.

The Thinking Systems in Historical Perspective

As an extension of this overview, we might also notice how these systems have timed their appearance as civilization has become more complex. We will develop this time line more fully in chapters five through ten, so permit us to touch it only briefly here.

The first four thinking systems have been influential for millennia, dating well back into ancient history. System 3 was already building empires more than 5000 years ago. And System 4 was ensconced throughout the Mediterranean centuries before Christ, giving us the philosophy of Greece and the theology of Israel.

Then, over an extended era, stretching through late antiquity and the Middle Ages, no new systems appeared. Complexity was constant enough that the first four systems provided adequate models until the earliest moments of the Renaissance. The birth of modern science changed all that. System 5 began to show itself in the thirteenth and fourteenth centuries, ascending to greater and greater prominence over the next 400 years. It empowered the Industrial Revolution and fueled the rise of Western capitalism.

By the late nineteenth and early twentieth centuries, complexity was advancing so fast that three more systems rapidly triggered. Two of these, Systems 7 and 8, have activated during the lifetime of today's oldest living Americans. And what of System 9? Is there such a thing? More than likely. Should complexity begin to overwhelm System 8 models, we will probably discover that God has hidden other thinking systems in the mind. We are not aware of them at present because complexity has not reached a level that calls for them. In that regard we are like observers in the nineteenth century, who had no way to foresee Systems 7 and 8, even though they were just around the corner.

4

Dominant System Transitions

As we have seen, dominant systems determine our core models. Through these models a dominant system embeds its distinguishing outlook and priorities in our behavior. In the parlance of psychology the dominant system constitutes a "deep structure" in the mind, governing our responses almost like instinct. We might describe a dominant system as

> a cogent neuro-psychological structure with which our values
> are arranged, prioritized, and interrelated, so that there is a
> general consistency in our response to stimulus or change.

We use the term the term "neuro-psychological" because there is evidence that these systems affect neural mechanisms in the brain.

For instance, every thinking system has a preference for certain values and outlooks. Dozens of words and phrases connote those preferences. (We will look at these extensively in chapter twenty.) Clinical tests have shown that our visual system is super-sensitized to words associated with our dominant system. We actually recognize those terms more rapidly than we do words connected with other systems.[1] It is hard to account for that phenomenon unless neural structures in the brain adjust to dominant system priorities.

It is because these systems embed themselves so deeply in the psyche that they begin to govern our behavior almost like instincts. Since social units bring birds of a feather together, groups tend to attract members who share the same dominant

---

[1] The National Values Center (P.O. Box 797, Denton, Texas 76202-0797) provides these test results in various handbooks for their seminars.

system. This shared outlook then becomes characteristic of the group itself. It is as though the dominant system has implanted itself in the "collective consciousness" of the group as well as the personal mindset of each member.

Businesses, churches, families, tribes, even entire societies bear a distinguishing system texture. You can "feel" it when you are in their midst. Once identified, that dominant system is an accurate predictor of how an association of people will organize itself, respond to crisis, and build interpersonal relations. We will treat this subject more fully in chapters five through ten. But to offer an illustrative example for the moment, we might think of two different congregations and their worship services. One of these churches has a high System 6 dominance. The other is dominant in System 4. The atmosphere in their worship will reflect that contrast. The System 6 church will emphasize spontaneity, interaction, and informality in its worship. The System 4 church will elevate structure, predictability, and minimal innovation. (And both, incidentally, are likely to defend their approach as the better exemplar of biblical worship.)

Notice that both these congregations may use the same *means* of worship (songs, prayers, Scripture readings), but they take a sharp departure from one another in their *mode* of worship. That contrast is no accident. When we adopt a new dominant system, it may not change *what* we do. But there will be a distinct change in *mode*, the *way* we do things.

Modes of Existence

To emphasize this change of mode from one system to another, we will refer to individual thinking systems in the future as "modalities." We have chosen this term because of its close semantic relationship to "mode" and "models." Each *modality* ushers in a different *mode* of existence, which is a reflection of that system's core *models*.

The term "modality" also highlights the fact that each system represents a *thinking mode* which people may share, whether they agree in their personal beliefs or not. Remember that conceptual systems do not determine *what* we think so much as the *manner* in which we think. To say this another way, thinking systems relate to the *structure* of our thought more than the *specifics* of it. For example, in ancient System 2 cultures people typically thought the same way, i.e., mythologically. But from one culture to the next the myths had their own unique content.

Or to select a more modern example, both Marxists and Christians believe that history has a predetermined outcome. This reflects a System 4 view of reality, which posits an inexorable guiding force behind human events. But System 4 does not presuppose, in and of itself, whether that force is personal or impersonal or where it may be taking humanity. Thus, when Marxists and Christians talk about human destiny, they present conflicting conclusions, for they are miles apart in *what* they think. In both instances, however, their mode of thinking — the *manner* in which they view historical processes — reflects a System 4 perspective.

Resistance to Changing Models

Dominant modalities do not mature overnight. After all, they represent new thinking patterns, and those patterns need time to become second nature to us. It may take years, perhaps decades, for a new dominant system to mature fully. In the case of cultures or societies the time frame can even span centuries.

One reason for this protracted process is that the mind has a built-in bias against changing cherished models. We often cling to models, even casual ones, in the face of striking evidence that contradicts them. Our minds accomplish this feat by using one of two techniques. We might call them the "shrug off" principle and the "twist it" principle.

The "shrug off" principle works like this: when we encounter something that does not fit our model, we tell ourselves that it only *appears* not to fit. That frees us to shrug it off. We are confident we could resolve the discrepancy if we understood this odd piece more fully. But not having time to work with oddities at the moment, we set it aside. We park it out of the way, along with other curiosities. Like Scarlet O'Hara in *Gone With the Wind* we vow to "think about that tomorrow."

A second stratagem, the "twist it" principle, allows us to distort data that violates our model. Like Cinderella's sisters trying on the glass slipper, we twist the offending reality to "make it fit." If I come to a traffic signal whose bottom light is blue, I try to make sense of that arrangement by "twisting" the incoming data stream. "There is a red light on top and a yellow light in the middle, but no green light," I say to myself. "Therefore, in this case blue is just a variant of green." Now the experience "fits."

Flashing Green Lights

Nevertheless, we sometimes come up against realities that can neither be shrugged off nor twisted to fit our model. When that happens, we begin to rethink the validity of the model itself. A drive through British Columbia once brought our family upon a traffic signal with a flashing green light. Our "traffic signal model" could manage a flashing red light. It also knew what to make of a flashing yellow. But it had no place for a light that flashed and was also green. Try as we might, none of us could force this new experience into our model.

Later we learned that the light was a warning. It alerted drivers to a pedestrian-controlled signal at a crosswalk in the middle of a block. Once we understood that convention, we modified our model to include flashing greens. In this case our model was so patently inadequate that we revised our model immediately.

In other instances our decision to revamp a model evolves more slowly. The change comes over a protracted period, as we twist more and more experiences to make them conform to the model's dictates. Like barnacles attaching themselves one by one to a ship until they finally impede its progress, these twisted realities start to impair the model's efficiency. Eventually we concede that the model is outworn and needs reengineering. Perceptions of Japanese products changed this way from 1960 to 1980. This is the same process by which we become fast friends over time with someone we previously disliked.

System-to-System Transitions

If we are reluctant to change even *casual* models, we are doubly reluctant to overhaul *core* models. As a result, we transition to new dominant modalities at something less than a uniform pace. Our progress is more nearly a stop-and-go affair. It is reminiscent of a line from Robert Frost. Thought, he said, flows like water spilled on dry ground — running, gathering, breaking forth, and running again. Dominant modalities develop in similar phases. There are times of "running," when we hurriedly accumulate a host of new models. Then, we slow down for a time of "gathering," to use Frost's term.

We need this respite to become accustomed to our new perspective on life. It is like adjusting to a first set of bifocals. We are thrilled at the improved vision they give us. But our mind is not used to a split field of vision in front of the eyes. Initially this way of looking at the world feels awkward, even a bit unsettling. With time, however, the mind reconditions itself, and we learn to manage the bifocals with grace and ease.

To borrow from Frost again, there may be several times of "gathering" before we complete the transition to a new dominant modality, for these transitions usually occur in incremental steps. As a result we spend much of our life, not anchored to a single dominant modality, but working our way from one to the next. We are like passengers traversing the early American West by stage. Each leg of the journey would end at a way-station, where the travelers might take an extended rest. Between stations, however, there was many a demanding mile.

Transition Zones

In much the same manner, we can view the modalities as way-stations, with protracted intervals between them. Figure 4-1 is one way to depict these transition zones. As we move through a transition we gradually subordinate core models from our former dominant system to those that are part of the emerging one. Notice, however, that we *subordinate* the previous models. We do not discard them. They continue to be available to us, even if somewhat in the background.

In other words, new dominant modalities enlarge or build on top of models they inherit from previous systems. Because those earlier models remain at our disposal, we can shift back to them if the occasion is appropriate. In the course of a week we typically shift from system to system dozens of times, often opting to spend hours in a non-dominant system.

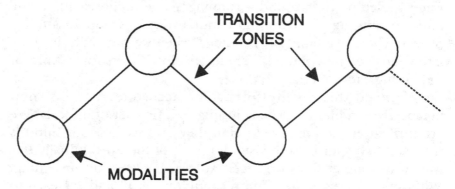

Figure 4-1: Transition Zones

To cite a simple example, when we take part in highly competitive, contact sports, we are likely to adopt a System 3 perspective, no matter what our dominant modality is. In System 3 winning and losing is everything. Its goal is to trounce the opponent. But however competitive we are on the field, once the game is over, we may return to a dominant system where we would never think of trouncing anyone.

Or we might go to Newtonian and Einsteinian physics for another example. The former emerged from System 4/System 5 thinking, the latter from System 6. Newton explained the circular motion of planets by postulating a force called gravity. Einstein's model accredits that motion to the curvature of space. If we look at the universe from Einstein's perspective, gravity in the Newtonian sense does not even exist.

Does that preclude anyone who accepts Einstein's view from ever talking about gravity? Of course not. In describing the trajectory of a basketball shot, gravity is an adequately complex concept. We do not need Einstein's model to account for the path the ball follows as it arcs through the air. On the other hand, Newton fails us when we try to explain other pathways that form an arc. One of those is the bending of radio signals as they pass near the sun. There we need the greater complexity of Einstein's model.

Borrowing from Systems Theory

By now you understand enough of the thinking systems basics that you are ready for the detailed system descriptions that we have promised. Before we turn to those introductions, however, permit us to address one other item of business. By referring to the modalities as "systems," we imply that the principles of systems theory apply to them. Yet, to this point we have not drawn out that implication. As we close this chapter, then, let us do so briefly.

What we have said thus far can be reduced to three principal concepts from classic systems theory. These are circularity of process, co-causality, and homeostasis. Since these will be new terms to some readers, allow us a moment to introduce them. First, circularity of process.

One characteristic of systems is that they link together processes so that some type of circulation occurs. We speak of water circulating through the cooling system of a car. Or air circulating through the heating system in our home. In effect, processes inside a system feed back on themselves. As a result of this circularity, co-causal relationships are set up. Co-causality, our second principle, simply means that two realities have the potential to change each other. The thermostat in your heating system has the potential for turning on the furnace. But the furnace, by raising room temperature, has the potential to trip the thermostat, which shuts the heat off. The furnace and the thermostat have a co-causal relationship.

When we examine the purpose for this circularity and co-causality in a system, we usually find that they work together to maintain some type of balance. An inventory system seeks to maintain an appropriate level of stock on the shelves. An air conditioning system aims at balancing the temperature at an assigned level.

The technical name for this balance-seeking is "homeostasis." It combines two Greek words that mean "to stand as one." To fully comprehend the dynamic inside a system, we must first determine the balance the system is trying to achieve. We also look at the interplay of various elements within the system, examining the co-causality between them.

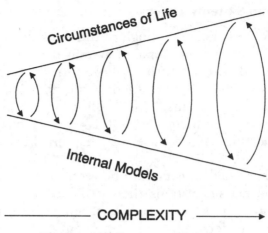

COMPLEXITY

Figure 4-2: Systems Dialogue

Figure 4-2 depicts how the three systems principles we just described evidence themselves as we move from one dominant system to another. The paired arrows in the graphic each represent one of the thinking systems. You see the circular pattern here as each of these systems maintains a dialogue between the external world (our life circumstances) and our internal models.

Now, place yourself inside the dialogue of one of the systems toward the left. What this dialogue aims to achieve is a balance between the conditions of your existence and the models you use to make sense of that existence. The arrow that points downward indicates the process of evaluating events by comparing them to internal models. The one pointing upward represents your response to those events, based on this evaluation.

There is, in other words, a co-causal relationship between the events that occur along the two lines labeled "Circumstances of Life" and "Internal Models." Each can change the other. External events can cause you to reevaluate or modify your models. But your models will determine how you seek to shape external events.

So long as your models are basically adequate to manage external complexities, you are in a state of homeostasis. You have no need to move to one of the modalities to the right. But as complexity increases (indicated by the line at the bottom of the drawing), life's realities start to overwhelm your core models. This upsets your homeostatic balance. Your cognitive and emotional systems then respond as systems always do by trying to restore homeostasis.

To do that, you will begin to revise your models. You will work to make them more comprehensive, capable of managing the more complex realities that have come into your life. Because your models are becoming more comprehensive, however, they are also becoming more complex, just as your life circumstances have already done.

And how long will you continue to modify core models? Until you reach a new balance point. You will eventually come to a place where your internal models are once more adequate to accommodate the complexity of your existence. If this revision takes you far enough, you will move into a new dominant system altogether. But your stopping point may actually be at some place in between two systems. These are the regions we identified as "transition zones" a moment ago.

There is a fourth systems principal at play here, what we call "synergism." But we can best describe its role later in our study. In chapter twelve, as we begin to see how these systems play off one another, we will return to Figure 4-2 and show how synergism is at work.

5

Systems 1 and 2: The Quest for Safety

In terms of conflict in today's church, Systems 4, 5, and 6 are the key players. Resolving the tension among them is a key objective in systems-sensitive leadership. To build a healthy congregation, however, we need to affirm and build health in all thinking systems, not just those that are currently at odds. Systems-sensitive leaders should understand all eight modalities, at least well enough to know their identifying marks, their typical patterns of behavior, and their distinctive ways of looking at the world. Chapters five through ten aim at providing that understanding.

We start in this chapter by examining Systems 1 and 2, which shaped the very earliest human societies. Chapters six and seven take us into Systems 3 and 4. System 3 built the legendary empires of the ancient world, then teamed with System 4 to govern the Middle Ages. As capitalism and modern scientific thought began to appear, System 5 joined them. We look at that modality in chapter eight. In chapter nine, we focus on System 6, followed by a quick look at Systems 7 and 8 in chapter ten. These last two modalities were all but unknown before the twentieth century. But they represent the thinking pattern of millions today and millions more tomorrow.

Some Initial Considerations

Before we begin our survey, some general observations are in order. First, anticipate that you may find a bit of yourself in most, if not in all of these systems. That is the result of pervasive mass communication. Broadcast and print media are not system-selective in the messages they transmit. They bombard us with concepts and outlooks from every modality. In today's world most well-read people are conversant with major terms and ideas from each of the systems. Beyond that, an occasional viewpoint sometimes strikes a resonant chord in us, even though it comes from a system far removed from our dominant modality. A system does not have to be dominant, therefore, for us to find it familiar.

Second, not only will you identify closely with some systems, less so with others, you may also discover you react negatively toward certain modalities. This adds another dimension to organizational conflict. When two people have different dominant systems, that fact alone invites tension. Tension only escalates if one person deeply dislikes the other's dominant modality. Additionally, it is possible to affirm a system and reject it simultaneously. This can occur even within our own dominant system. Despite our preference for that modality as our principal thinking system and our reliance on it for core models, there may be things in it that grate on us. Organizations, too, can have a dominant system they both accept and disdain.

Third, every system has specific strengths and specific weaknesses. We will touch on both of those as we proceed. New modalities tend to emerge to counter inadequacies or excesses in the present dominant system. In the system descriptions that follow, therefore, we will point out shortcomings in every modality, not out of disrespect, but to show how its downsides cause other systems to trigger.

And finally, be aware that systems often manifest themselves more revealingly in organizations or societies than in the psychological profile of a single individual. Consequently, as we describe each system, we will look at how it patterns itself both at a personal level and across tribes, nations, and corporate organizations. Many of our examples will come from biblical and secular history.

At first you may question the value of discussing how Romans used System 2 to cement an empire or how Americans used System 4 to fight the Cold War. On the surface such subjects have little apparent relevance to your church. But this historical approach turns out to be an excellent way to uncover the inner dynamic of individual systems. The interests, issues, and tendencies of a modality are the same, whether we are dealing with that system in large sociological units (like a nation) or small ones (like a church).

As a consequence, what you learn about systems from historical and cultural examples will translate directly into congregational life. It may actually be preferable, indeed, to use a frame of reference *outside* the church to learn about systems dynamics. All of us are emotionally engaged with the systems inside our own church. We are so close to those systems that we may be unable to see them clearly. By learning to recognize thinking systems in other settings, we become more adept at identifying them in congregational contexts.

System 1

Our starting place is System 1, not just for this survey of system characteristics, but in life itself. System 1 is active the moment we are born. Physical survival is almost its exclusive concern. Values in System 1 prioritize around physiological needs: nourishment, elimination, warmth, physical comfort, etc. Consequently, life is a stream of short-term reactions and the world is perceived in kinesthetic categories (feelings). Do I feel hungry? Do I feel cold? Am I thirsty? In short, System 1 equates well-being with satisfaction of physiological drives.

Not only do we begin life in System 1, we revisit it briefly from time to time. When firestorms work their way along the California coast, people caught in the immediate path of those wildfires have a singular focus, getting out alive. Every life-threatening situation thus brings System 1 to the fore, if only momentarily.

In the case of protracted terminal illness, the semi-invalid condition of the final weeks marks another period of System 1 existence. The entire day's schedule revolves around how we feel. Activity is reduced to minimal nourishment and efforts to ease pain. Some homeless people also function in System 1, living one day to the next, wandering wherever the prospect of food or shelter takes them.

Time is relatively meaningless in System 1, for life is lived in the immediate moment. Nor is there a sense of "turf," since System 1 has no use for that notion. After all, infants (who are the largest segment of the world's System 1 population) have no control over where they are taken. And those who return to System 1 later in life are always in circumstances that make staying alive the only preoccupation.

Life at the Bare Edge

Anytime sheer survival is at stake, System 1 issues will frequently be prominent, even though other modalities may also be active. We saw this in the concentration camps of the Third Reich and the gulags of Stalinist Russia, where people often lived at the bare edge. Aleksander Sohlzhenitsyn captured the essence of System 1 thinking on the final page of *One Day In The Life Of Ivan Denisovich*, the story of a political prisoner in a notorious gulag. Lying on his bunk at lights-out, Denisovich recounted the last few hours in his mind.

> A lot of good things had happened that day. He hadn't been thrown in the hole. The gang hadn't been dragged off to Sotsgorodok. He'd swiped the extra gruel at dinnertime . . . . He hadn't been caught with the blade at the search point.

Sohlzhenitsyn concluded, "The end of an unclouded day. Almost a happy one." Only System 1 could think such minimal existence "almost happy."

Not just individuals, but entire societies are sometimes reduced to System 1 existence. The horn of Africa, assaulted with one devastating drought after another in recent decades, has produced circumstances where millions of people live in System 1. They wander from relief camp to relief camp, sitting for hours with the same empty stares we remember from concentration camps. System 1 knows little about hope.

As a result, in societies where this modality prevails, there is virtually no leadership or organizational impetus. People seem to do things with no design or structure. With nothing to hope *for*, there is nothing to work *toward*. It could even be argued that there is no such thing as a System 1 society. When circumstances become so dire that an entire populace plunges into System 1 survival tactics, nothing truly called a "society" still exists.

In terms of congregational dynamics, System 1 is a peripheral system. It is the dominant modality only for those who are either too young or too incapacitated to be active in the church's daily life. In many ways we could omit System 1 from our study since it has so little spiritual or organizational thrust. But to do so would be a mistake. Understanding System 1 is a necessary backdrop for comprehending the inner workings of System 2, which *does* figure significantly in religious experience.

System 2

Life in System 1 is a precarious one. We are utterly helpless against external forces that buffet us daily. We have no control over when we will eat next, what we will wear tomorrow, or even if there will be a tomorrow. There is no way to ignore our total vulnerability, even for a moment. Once we emerge from that uncertain existence, we still cannot let our guard down. We know that the slightest disruption could plunge us into survival throes once more.

It is only natural, therefore, that we begin to speculate about what we experienced in System 1. Why was our existence so unpredictable? Was some unseen power controlling the forces against which we struggled? If so, how can that reality be placated, its friendship secured? With these questions, System 2 is activating. In System 2 the life of reaction that dominates System 1 gives way to a life of vivid imagination. System 2 uses imagination to envision the unseen in an effort to comprehend it.

For children, envisioning the unseen means an intriguing world of fantasy. As they begin the transition from System 1 to System 2, fairy tales start to fascinate them and imaginary friends become their playmates. They draw heroes and heroines from the make-believe world of "once upon a time." This is the age of Santa Claus, the Easter Bunny, and the tooth fairy.

But not everything children imagine is benign. There may be monsters in the darkness or scary things under their beds. By imagining this admixture of good and bad creatures, the child is working through System 2's anxiety that perhaps the unseen forces are not friendly, but hostile.

As we mature, System 2 continues to struggle with this same anxiety, although no longer in childish ways. Why is there evil in the world? Why do bad things happen to good people? When I pray, why don't I get answers? Each of these is a System 2 question. Because such concerns are so vital to faith, all religions, even the most advanced, retain strong System 2 attributes. At the heart of System 2 is the sense that the unseen world is girded in mystery and marvel. Thus, awe grows out of System 2, as does the concept of inner communion with a higher power.

Placating Unseen Powers

Not surprisingly, societies that function in this modality entrust power to those with "spiritual" prowess. Shamans, witch doctors, priests, and priest-kings control the life of the community. In System 2 the most prominent feature of religion is sacrifice and ritual. Religious activity centers on those who are adept at ceremonies to please the gods. Almost universally (Judaism in its System 2 phase was a notable exception) the purpose of sacrifice is to appease or "buy off" deity. This preoccupation with manipulating the spirit world easily degenerates into magic and superstition. System 2 also gives rise to horoscopes and palm readers, ouija boards and tarot cards.

In primitive forms System 2 religion often slips into animism, largely because System 2 is not capable of sophisticated abstract thought. Feeling is as important to System 2 as thinking. When it does think, System 2 depends either on symbols to express the ineffable or concrete terms to explain personal experience. Genuine abilities at abstraction will become prominent only when System 4 activates.

As a result, System 2 may not distinguish between the universe and the power behind it. Pantheism is one result. Another is the worship of rivers, trees, the sun, the moon, mountains, and even rocks. System 2 often views everything as having spirit or personality within it. If a stone hurts me, something in the stone must not like me. If a river floods our home, the river must be angry at us.

Suspending Time and Reality

Small children talk about their dolls and toy animals in much the same way, as though these playthings were alive. The comic strip *Calvin and Hobbes* has capitalized on System 2 dialogue between a young boy and his stuffed tiger. They argue, fight, and frolic together. Their interaction shows System 2's tendency to blur perceptual lines between the imaginary and the real.

We must not be condescending about that quality, however, for we draw on it often as adults. Every time we become caught up in a movie or find ourselves drawn into the pathos of a drama, we are tapping System 2's ability to suspend the distinction between reality and imagination.

Interestingly, if we become enthralled by a performance on stage, time all but evaporates. This, too, is characteristic of System 2, where little sense of time exists. To the extent that System 2 thinks of time at all, it views it as a cycle of special events. For the child these include birthdays, Christmas, and vacations. In System 2 societies the cycle centers on holy days, festivals, and seasons for planting and harvest. Our church calendar, with its cycle of Lord's Days and its annual recurrence of Easter (as well as other holy seasons in many denominations), reflects the continuing role that System 2 plays in shaping our religion.

A Place to Feel Secure

System 2 also gives rise to our first "turf awareness," which basically translates into the idea of "my safe place." For children it is the notion of "home." For System 2 societies it is "our sacred place." They feel a sense of security at a hallowed spot, for they presume that it enjoys divine protection. After all, the gods must surely have a vested interest in the site where they receive sacrifice. Any harm done someone there would clearly invite the wrath of deity.

In the ancient world this notion of a protective divine presence led to the inviolate nature of temples. No one dared desecrate a holy place by slaying someone who stood within it. Famous stories in both Jewish and Greco-Roman culture turn on that principle. Nor did this concept disappear with antiquity. Well into modern times both custom and law precluded arrests inside a church building. Being a "holy structure," it offered sanctuary for those within.

Unless we comprehend the compelling need for "existential safety" within System 2, we will not understand its underlying dynamic. Because System 2 has only a rudimentary idea of "society," it has no perception of social threats. Endangerment comes only from specific individuals and nature.

Therefore, System 2 looks for an alliance with forces who can keep evil people and evil things at bay. Young children in modern societies often say they want to be policemen, firemen, or doctors when they grow up. Those professions attract the child because they symbolize a defense against bad people and bad things. "If the policeman, the fireman, and the doctor are on my side," the child reasons, "everything will be okay."

Ritual and Repetition

Because security against the unseen is such a compelling concern, System 2 is drawn to do things in ways that have proven successful in the past. Since the elders (or in the case of children, the parents) are authorities on the past, we defer judgment to them. The tribe's way is the only way. Children are making this very point when they say, "That's not how my mommy does it."

Security within System 2 casts itself in repetition, ritual, and totems. If the way our forefathers did things worked for them, we must be careful to follow an identical pattern in the future. To be certain we have captured every nuance of those rituals, we repeat them over and over until they are indelibly committed to memory. This repetition, especially as it occurs in primitive dances and music, strikes the modern mind as monotonous. But to System 2, repetition and routine are reassuring.

That is why small children ask for the same bedtime story again and again; why they want us to read them the same book night after night; why they wear out a favorite video, watching it over and over. This need for ritual also underlies the uneasiness many children feel if they do not have predictable bedtime routines. A consistent pattern of events, repeated each evening, offers a final reassurance before entering the uncertain realm of darkness.

Tribe-Building

In terms of identity, System 2 builds a sense of "my tribe," which equates with "those who gather at my safe place." To the child, family becomes the tribe. In a society dominated by System 2 thinking the tribe consists of those who share the same holy place. In System 2, personal identity is largely inseparable from "tribal" identity. I become "one" with those who are fellow-members of my tribe. The Lord's Supper itself addresses this inner need to find "oneness" with the "tribe," in this case the community of believers. As Paul says, "We who are many are one body because we break one bread" (1 Corinthians 10:17).

In that statement the apostle touches on a key feature of System 2 values. To be specific, System 2 employs ritual, ceremony, and symbol to secure psychic unity in the tribe. Families develop their sense of tribe with such things as pre-meal prayers, Thanksgiving and Christmas traditions repeated each year, and the special way they celebrate birthdays. The regularity of these rituals offers assurance that the tribe is still functioning. Singing "the old songs" serves a similar function in churches, as does a certain consistency in the order of worship from week to week.

Entire nations use rituals to create a sense of tribe. When children begin their schoolday with the pledge of allegiance, or when the band plays the national anthem at pre-game ceremonies, we are celebrating tribe-building rituals. The monuments along the mall in Washington, D.C. are nationalism's equivalent of "our holy place." These memorials embody the "hallowed memories" that make us one people. Patriotic music stirs up warm System 2 feelings in us, as does a July 4th parade or the picture of a President's flag-draped casket. Thus, while we think of primitive societies as prone to ceremony and ritual, even the most advanced cultures need the tribal cohesion which System 2 affords.

Play in System 2 is also a tribe-building experience. Kicking a ball in the backyard with Dad does not seem to be a game if we think of games as something in which one wins, another loses. But in System 2 the object of games is to build a bond between ourselves and those with whom we play. Games are therefore collaborative, not competitive. A family working together on a jigsaw puzzle is playing a System 2 game. So is a youth group painting a mural in the church annex.

The Sense of Taboo

In a sense, System 2 never loses its childlikeness. There is no competition with others, no preoccupation with status, no premeditated violence or greed. Everything is to be shared, and the world is a joyful place to explore. It is as though System 2 lives in a state of suspended innocence. Sociologists speak to this point when they describe an almost childlike naiveté that governs many primitive people. Within System 2 there is no sense of guilt as such. A true concept of guilt comes with System 4. What System 2 knows is taboo, which we communicate to children when we tell them something is "dirty." To children the parents serve as sources of authority on what is taboo, just as elders of the tribe provide that authority in System 2 societies.

Violation of taboo always induces a sense of defilement or uncleanness. To remove that "dirtiness," System 2 religion emphasizes purification rituals. These may include washings, blood sacrifices, or some rigorous ceremony aimed at purging oneself of whatever defiles. Judaism, with its extensive codes of ceremonial uncleanness, maintained a vibrant System 2 strata. And the emphasis on personal purity in Christianity, not to mention the purifying work of Jesus, are extensions of System 2 themes.

System 2 Culture in Scripture

As you think about our description of System 2, you can quickly identify numerous scenes in the Bible where motifs from this modality abound. The stories in Genesis evidence heavy System 2 shading. Jacob's vision of the ladder into heaven. His consecration of Bethel as a holy place. His wrestling with God on the banks of the Jordan. Or earlier, Abraham's sacrifice of Isaac. And the night Abraham saw a flaming torch and a smoking pot pass between the pieces of his sacrifice.

These and many other stories in the Pentateuch are latent with System 2 meaning. They speak of an era in which the forefathers of Israel, living in a society dominated by System 2 thought, were largely System 2 thinkers themselves. These stories strike us as extraordinary because our dominant modalities are so far removed from System 2. While System 2 still touches certain aspects of our lives, it is not our fundamental method of conceiving the world. With Abraham, Isaac, and Jacob, by contrast, it was their primary conceptual system.

6

System 3: The Quest for Power

When small, once-isolated tribes find themselves competing with others for water supplies, grazing rights, or
arable soil, a new set of existence issues emerges. No longer can
the tribe wander wherever it wishes, without opposition or
hindrance. Faced with the exploitive intent of powerful neighbors,
the resources of System 2 prove inadequate to the challenge. To
overcome that deficiency, society begins a transition to System 3.
When historians talk about the "dawn of civilization," they usually
are referring to the first stirrings of System 3 in human history.

Existentially small children go through a similar phase when
they start contending with playground bullies. The rules of play
that worked in the safe, protective shelter of home no longer
suffice. This is a world that threatens to push us around, and will
do so if we do not learn to push back. The time has come to fend
for oneself.

Whereas System 2 concerns itself with threats from the unseen
world, System 3 perceives its primary threat as coming from other
humans. That change in perception leads to a radical redefinition
of security. Ritual and tribe-building are powerless against this
new danger. What our circumstances now call for is raw physical
power to vanquish any adversary.

A Power Ethos

System 3 lives by the rule of might. In the ethics of System 3
the strong survive and the weak perish. Winners make the rules,
not by virtue of moral or intellectual superiority, but simply

because they have prevailed in the struggle. Losers have no voice because they have no power. Might makes right.

With this hunger for power, System 3 develops an insatiable appetite for conquest. While System 2 builds tribes, System 3 builds empires. Assyria, Babylon, Greece, and Rome were forged on a System 3 anvil. So were the feudal systems of Europe, Asia, and the Middle East. System 3 later carved out colonial empires in the New World and conquered the American West. All imperialism, indeed, grows out of this system, as does the "dog-eat-dog" business of raw marketplace competition.

To maximize its strength, System 3 abandons the System 2 view of nature. System 2 tries to live in harmony with nature, fitting naturally into its flow and marveling at its mysteries. System 3, on the other hand, does battle with nature, wresting power from her by exploiting her natural resources. System 3 sees nature as just another force to subdue in the course of building might.

Warlords and Empires

In the transition from System 2 to System 3, priest-kings give way to warlords. The issue at hand is no longer the need to placate deity, but to defeat the enemy. System 3 invariably supersedes leaders who know rituals with warriors who know tactics. In biblical history that moment came when Israel no longer wanted Samuel to lead them. Instead, they asked for a man like Saul who could give them an army (1 Samuel 8:4-9).

Moreover, System 3 basks in self-glorification. Its empires erect impressive monuments to celebrate their triumphs, dazzling capitals to show off their splendor. With this passion for building on a colossal scale, System 3 has achieved magnificent breakthroughs in architecture, engineering, and art. Of particular note have been its magnificent temples, like those that lined the Capitoline Hill in Rome. Or the Parthenon on the Acropolis in Athens. System 3 accredits the gods with giving the empire success. System 3 reciprocates by housing these gods in a manner befitting imperial glory.

You will immediately recognize that David and Solomon followed a System 3 script in securing an empire for Israel. As soon as David consolidated his reign, his first priority was to build a resplendent capital in Jerusalem.[1] Solomon then expanded the empire, and with that expansion, exceeded his father as a builder. Both men, of course, are remembered for their grandest endeavor, the temple on Mount Zion. The language of 2 Samuel 7 implies that this project was David's idea, not God's, further evidence of David's System 3 stripes.

Campaign Seasons

With its militaristic thrust, System 3 develops a new sense of time. Rather than base a calendar on holy seasons and cycles of religious festivals (as System 2 does), System 3 organizes time around campaign seasons. In the ancient and medieval world those campaigns were military in nature. Today we talk about campaign seasons in politics or pennant campaigns in baseball. When we do, our vocabulary bespeaks a System 3 outlook on time. As soon as the Super Bowl is over, football enthusiasts start looking to "next year," as though nothing important will happen before then. Not only that, "next year" will actually be "this year" in terms of the calendar. The Super Bowl is in January, the opening of season in late summer. But the System 3 mind thinks only of a campaign calendar, not the actual passage of time.

Power Pyramids

Once we recognize the strong militancy in System 3, we are not surprised that it is the first system to develop a distinct pecking order. It builds a hierarchy of emperors, princes, dukes, and counts, something unknown in the more egalitarian world of System 2. System 3 gives us the tough shop foreman who runs things with an iron fist and cannot be intimidated. System 3 also drives the "don't-give-me-any-excuses" manager who whips a troubled organization in shape by brooking no indolence or insubordination.

---

[1] So successful was he in this endeavor that Jerusalem became known as "the city of David." Beginning in 2 Samuel 5:7, the Old Testament uses this name for Jerusalem 40 times, but never (as Luke 2 does) for Bethlehem.

System 3 organizations are always in the form of a pyramid, with a tiny elite at the top. This leaves the commoners at the bottom of the heap. Usually powerful elites hold them there in forced subjugation. On occasion, however, commoners acquiesce to the pyramid structure because they believe it affords security from predatory enemies. In exchange for safety they accept a voiceless status. Employees may put up with a dictatorial new boss if they believe he is the last chance to turn the business around and save their jobs.

The original System 3 pyramids of the ancient world were aristocracies based on military titles and achievements. When military necessity no longer justifies that type of social structure, System 3 stays in control by stratifying society along other lines, most commonly economic or racial ones. System 3 demands that everyone "know his place" and keep to that place in the hierarchy. In the early 1970s, the award-winning Broadway musical *Purlie* looked at this type of social structure in the deep South. The play's central character was a black preacher whose congregation of sharecroppers lived under the economic despotism of Ol' White Cap'n, the county's principal landowner. White Cap'n justified his merciless manner by reminding everyone that big fish eat little fish and little fish eat the littlest fish. If nature decreed that order, who was he to argue?

White Cap'n's outlook embodies one of the strongest tendencies in System 3, namely, its willingness to be unabashedly exploitive. Historically one of the first things System 3 exploits is System 2. To carry out its own aggressive ambitions, System 3 needs clan solidarity, a readiness to fight as one. That makes System 3 dependent on the tribe-building rituals of System 2.

Thus, organizations that are System 3 dominant draw on System 2 tribe-building to induct new members into their ranks. The purpose of these System 2 inductions is to create psychic bonding within the group. Hazing in college fraternities is part of this process. Military boot camp (which builds a new sense of tribe and instructs the recruit in "the ways of the elders") is another. Even street gangs have their own rites of passage, some of them quite violent, by which aspirants prove themselves worthy to be part of the tribe.

System 2/System 3 Alliances

Historically, when System 3 becomes dominant at a societal level, it does not banish System 2, but instead forms a strong coalition with it. System 3 moves quickly, therefore, to place System 2 leaders at its side. Sometimes System 3 does this sincerely (as when David made Samuel his confidant), at other times as a sheer political ploy. But whatever the motive, the pattern repeats itself endlessly throughout history. Indeed, System 3 often co-opts System 2 so fully that the political and religious aristocracy both come from the same families. When Julius Caesar seized the reins of government, one of his first acts was to name himself *pontifex maximus* (high priest) of the state religion. In the Middle Ages firstborn sons of nobility inherited their father's political title and the lands that went with it. The remaining sons often entered the clergy and rose to be bishops, cardinals, and popes.

The Romans were particularly astute at combining System 2 tribe-building with System 3 empire-building. They never tried to destroy the religion of nations they conquered. Instead, they incorporated the deities of those nations into the Roman pantheon. The Middle Ages provide another textbook study of blending System 2 with System 3, as church-state partnerships wove their way through the entire fabric of medieval society. And this was as true of Islamic lands as Christian ones.

The alliance between System 2 and System 3 has been a particularly resilient form of social structure. Its success over the centuries has given it staying power. In the twentieth century some of the ugliest forms of totalitarianism have used a convergence of System 2 and System 3 to maintain repressive regimes. In Germany a large-scale resurrection of System 2 in the form of mythology and occultism prepared the soil from which Nazi fanaticism grew. In Russia, where atheism was state policy, the Soviets used the quasi-deification of Lenin as a substitute System 2 religion.

All of which is to say that System 3 has a long history of manipulating religion for ulterior purposes. The Jonestown cult in Guyana and the David Koresh compound in Texas are modern-day examples. When it comes to religion, moreover, System 3 will exploit System 4 just as quickly as it does System 2. System 3 cloaked the divine right of kings in System 4 theology, just as it later justified American imperialism by appealing to manifest destiny, another System 4 concept. When Senator Albert Beveridge, a towering political figure at the turn of the century, defended Admiral Dewey's seizure of the Philippines, he called it "the work of the unseen hand of God."

System 3 even refashions God into its own warlike image. In the movie *The Longest Day*, the story of the Normandy invasion, General Eisenhower responds to an unpromising weather report by musing, "I sometimes wonder whose side God is on in this war." A few moments later the scene shifts to a German bunker on the French coast. Peering through binoculars into an impenetrable fog that might mask an Allied attack, the German commander says, "I sometimes wonder whose side God is on in this war."

Ruthless Violence

Despite its frequent imprecations of deity, System 3 lacks a moral code to temper its aggressiveness. In developing its own ethic, System 3 gains meager guidance from Systems 1 and 2, the only systems that preceded it. The sole value in System 1 is survival. And while System 2 does have a sense of taboo, it lacks a developed ethical structure (in part because System 2 has only a rudimentary concept of society). Thus, System 3 has little to divert it from a doctrine of might makes right. The strongest make the rules, and everyone else yields to them. But since there is no guilt in System 3, its rules are often ruthless. We see this moral insensitivity in the torture that children inflict on animals. Or again in the cruel taunts they hurl at one another. We observe this viciousness on an even larger scale in the unconscionable atrocities that System 3 societies perpetrate on those they conquer.

To curb System 3, we must meet force with force. This principle holds true whether we are dealing with a defiant nine year-old, urban street gangs, Mafia crime bosses, or Iraqi aggression in Kuwait. Because System 3 lacks guilt, reasoning with it has little effect. Neither does an appeal to humanitarian instincts. System 3 interprets such overtures as evidence of weakness. Hungry for power itself, System 3 only respects power.

Saving Face

If System 3's anti-social behavior becomes deeply ensconced, no amount of force — even overwhelming force — may suffice to intimidate it. In its most extreme forms, System 3 adopts the motto, "Better to die than to back down." Backing down would mean loss of face, and System 3 always tries to save face. It cringes at the thought of being humiliated. More than anything else, System 3 wants to be respected. Humiliation denies it that respect. Without the tempering of System 4, System 3 has a highly developed sense of shame with little counterbalancing sense of guilt. As a consequence, System 3 can shed any semblance of principle in order to save face.

Newsweek magazine chronicled this aspect of System 3 thinking in a May 9, 1994 article on American urban gangs. The story drew on an interview with Elijah Anderson, a sociologist who works gang-infested neighborhoods of Philadelphia. He speaks of the "code of the streets," at whose heart he finds "a pathological reverence for respect." Quoting from *Newsweek*:

> The craving for respect, which turns into a thin-skinned quest to prove "manhood," inevitably leads to violence. While others may walk away from a slight, street youths are required under the code to show their "nerve" — by pulling a trigger or throwing a punch. "Many feel that it is acceptable to risk dying over the principle of respect," says Anderson. "In fact, among the hard-core street-oriented, the clear risk of violent death may be preferable to being 'dissed' by another." It is also assumed that everyone understands the code. So if a victim of a mugging responds "wrong" according to the code — by, for instance, maintaining eye contact too long — the perpetrator "may feel justified even in killing him."[2]

System 3 often rationalizes such extremes by viewing its victims as something less than human. War propaganda is notorious for demonizing the enemy. Once we cast the adversary as subhuman, we can excuse whatever it takes to destroy him. We see a similar "dehumanization" in slavery. The Greeks referred to their slaves as "talking tools." And American slaveowners some-

---

[2] "Learning from the Streets," *Newsweek*, May 9, 1994, p. 61.

times argued that blacks had no souls.

In its harshest form, System 3 never sees itself at fault, even when it commits atrocities. System 3 views its problems as coming entirely from external sources. If things go wrong, someone else is to blame. In a nearby city a street tough killed a five year old during a drive-by shooting. Later, interviewed in prison, he shrugged off the crime. "She shouldn't have been sitting on that porch," he said. System 3 waves a banner emblazoned with the words, "Not My Fault."

Powerful, But Insecure

Ironically, the brute power of societies where System 3 is dominant ultimately becomes their undoing. The fundamental amorality of System 3 invites an escalating scale of heartless violence. In the process System 3 creates embittered enemies. Whenever the opportunity affords itself, the victims of System 3's ruthlessness retaliate in kind. Ultimately System 3 aristocracies and oligarchies find themselves living a precarious and insecure existence. Surrounded by hostile forces, both inside their power structure and outside it, they cannot let their guard down, even for a moment. The history of System 3 empires is written in the blood of palace revolution and the refined art of coup d'etat.

The amorality of System 3 also leads to an impulsive and hedonistic lifestyle. Since ethical sensitivity will not develop until System 4 matures, pleasure is the only feedback system to guide System 3 decision-making. If something is enjoyable, System 3 sees no reason to abandon it. If it is unenjoyable, System 3 has no motivation to see it through. System 3 has fun for the moment, with no real concern for long-term consequences.

System 3 is also ostentatious. It loves to show off the spoils of its hard-won victories. The pimp with his stylish clothes and expensive cars, the crack dealer with his gaudy rings and gold necklaces, the star quarterback with the gorgeous cheerleader on his arm — all are cases of System 3 sporting its trophies. So, too, are the overdone mansions built by the *nouveaux riches* who made their fortune in the tough System 3 world of mining, shipping, and heavy industry.

The undisciplined self-centeredness of System 3 in childhood can be annoying, even frustrating for parents and teachers. But it rarely causes alarm. If System 3 remains unchecked into adolescence and adulthood, however, its anti-social behavior and irresponsibility become deeply problematic. As it increases its independence from the family-tribe, System 3 becomes totally absorbed in self and self-gratification. The "macho male" is a stereotypical result of this self-centered, power-driven approach to life. So, too, is the *Playboy* lifestyle, which sees women as playmates of the month to be subjugated and exploited, then casually discarded for the next conquest.

What System 3 Contributes

On the other hand, when armed with a moral compass (which comes from System 4), System 3's aggression can be put to good use. We need System 3 to fight injustice, to marshal forces against evil, to provide us a competitive edge in business. System 3 affords us the enjoyment of organized sports and other forms of recreation in which the object is winning. System 3 gives armies their fighting form and communities their will to stand against oppression. Directed into proper channels, System 3 is powerfully invigorating for both individuals and societies. There are times that call for a firm, but benevolent dictator to pull things together and save the day.

System 3 is so important to personal development that it is one of the earliest modalities to activate. It gives us the tenacity we need to stand up to those playground bullies, to fight for ourselves, to take the initiative against adversity. In addition, the competitiveness of System 3 helps children work out their sense of identity. About the time this system turns on, youngsters realize that they have a future separate and apart from their family. Struggling with that idea, they become preoccupied with the question of who they are in their own right. They are eager to determine such things as, "What am I good at? What are my strengths? How well do I measure up against others?" In the push-and-shove world of System 3, they get quick feedback to help them answer those questions.

We can also view this phase of childhood as an unfolding recognition that "I can make an impact on the world around me." During the years when System 2 is dominant, children live in a state of passivity, content with what has been given them, fantasizing about other worlds, but quite happy with the one they have. As System 3 starts to rise, the child begins to wonder, "If I can imagine a different world, can I create a different world?"

Heroes thus become men and women of action. With boys especially, bedroom walls sprout posters of all-star running backs, all-pro forwards, or media-inspired super-heroes. For both boys and girls the "let's pretend" games of System 2 take a back seat to games in which scores are kept. These scores serve to establish the pecking order of who is best at what. Even under the guise of team sports, System 3 avidly tracks individual statistics to see who has the most field goals, the highest batting average, or the greatest total yardage. Playmates are no longer partners for forays into fantasy, but adversaries against whom we test our skills. In System 3 families the rite of passage for young boys is to prove their manhood through feats of courage and toughness. They learn to wear their battle scars proudly, as a badge of honor.

Manual Learners

As we have seen, System 3 solves problems by literally taking them into its own hands. Ironically, it also typically learns best by doing the same thing. As System 3 begins to develop in children, they start tearing things apart and putting them back together to see how they work. They become serious about the dexterity required to excel in athletics, music, and dance. System 3 societies learn how to achieve engineering marvels by massive projects of trial and error.

System 3 cultures therefore have little need for formal education. Children are apprenticed in a trade shortly after they are able to dress themselves without adult intervention. System 3, indeed, may sneer at people with "book learning." One of the great challenges in lowering the school dropout rate among inner city youth today is overcoming the System 3 code of the streets which sees the classroom as a waste of time.

Essential, Yet Dangerous

In summary, System 3 is absolutely essential to personal and cultural achievement. At the same time, it is potentially more dangerous than any other system. We have underscored its anti-social tendencies because congregational conflict has an uncanny potential to trigger this modality. Church fights can be as unprincipled as any other if factions revert to System 3 tactics. Partisans twist Scripture out of context, hurl insults, slander other Christians, and impugn motives. It can become quite an ugly scene. In the resulting crossfire the first victims are truth, love, and unity.

Channeled constructively, however, System 3 empowers the church to maintain the battle against spiritual enslavement and tyranny. Countless hymns, many of them age-old favorites, celebrate that principle. *Onward, Christian Soldiers. Soldiers of Christ, Arise. My Soul, Be On Thy Guard. Faith Is The Victory.* What a legacy of stirring music System 3 has given the faith! To the degree the church takes spiritual warfare seriously, such hymns will always hold appeal. To the degree the church envisions herself arrayed against hosts of unrighteousness, System 3 will continue to be essential.

7

System 4: The Quest for Truth

Some three thousand years ago System 4 began to add its influence to human affairs. Moses introduced System 4 thinking to Israel. The city-states of Greece and the republicanism of early Rome did the same for the Mediterranean world. By the time of late antiquity, System 4 had displayed its intellectual prowess in three monumental achievements — the codification of Roman law, extensive development of Christian theology, and completion of the Jewish Talmud.

For the next thousand years System 3 and System 4 worked shoulder to shoulder in an uneasy alliance, although System 3 dominated the political and social landscape. When imperial Rome collapsed, System 3 found new life in the rigid feudal privileges of medieval Europe. In the realm of human thought, however, System 4 was holding its own, toiling away in the monasteries and later the universities of both England and the Continent. There, in these isolated settings, thinkers hammered out ideas that would ultimately unseat System 3 power.

Toppling System 3

That moment came with the dawn of the early modern era. No single event gave System 4 its decisive victory. But the clash between constitutional democracy and the divine right of kings was clearly a watershed moment. By prevailing in that struggle, System 4 ensured its towering influence for the centuries to come. System 3 would not go away, of course. It is still very much with us, fighting our wars and fueling our competitive drive. But as a social phenomenon, it serves today only with the permission of later systems.

System 4 infuses a high moral sense into human affairs. It gives us our sense of absolutes, the commitments we consider nonnegotiable. It thus provides an ethical framework to temper System 3's impulsiveness and to curtail the excesses of later systems. System 4 works tirelessly to define principles and elaborate on their implications. Both of these skills require high degrees of abstract thought. For that reason System 4 does not blossom until a child's mind is capable of complex abstraction. Within societies System 4 becomes prominent only when prosperity gives rise to a leisure class that can afford the luxury of intellectual pursuit.

Disillusionment in System 3

Long before that cultural moment, however, System 4 makes its initial appearance. It usually nudges its way into the picture about the time that System 3 recognizes the folly of its violence. Early in the going, System 3 thinks itself indestructible, invincible. But protracted carnage and pain ultimately take the edge off that cock-sure attitude. One day, out of its peripheral vision, System 3 catches a glimpse of its own mortality. For street gangs in Los Angeles that moment came in the wake of the Rodney King riots. A few days after the turmoil, rival gang leaders held a joint news conference to declare a truce. "Instead of securing our turf," they explained, "our wars are killing our own families and neighborhoods."

Of course, their call for peace hardly signaled the end of gang violence. But it was a start. The System 3 code of conduct eventually leaves its adherents disillusioned. They come to see ruthless competition (whether between nations, businesses, political groups, or what have you) as yielding little true security. Instead of making life more secure, unbridled competition savages life and creates determined enemies, eager to repay System 3 in kind.

With that recognition, System 3 begins to seek some alternative to violence as a means of settling differences. But where shall it turn? The only alternative, it seems, is for warring parties to abide by mutually respected principles. The effort to enunciate those principles, in turn, moves us to a place where System 4 excels.

Truth and Transcendent Values

System 4 promotes harmony on the basis of shared beliefs. In place of System 3's passion for power, System 4 substitutes a hunger for truth. It seeks out ideas which are so compelling that everyone can unite behind them. System 4 therefore pursues knowledge with ardor. It builds great educational systems. It honors intellect the way that System 3 honors valor. System 4 is a literate, book-producing modality, for the printed page serves as a repository of wisdom, truth, and insight. This stands in contrast to System 2, which seeks knowledge at the foot of tribal elders. As System 4 matures, those elders lose their last vestige of power. System 4 no longer consults them. When System 4 needs information, it goes to a library.

In its quest for truth, System 4 treats one principle as self-evident — the sanctity of human life. To counter System 3's violence and callous disregard for others, System 4 demands rigorous respect for the human being. In System 4 every individual is important, a creature of worth and dignity. Because of this towering regard for the value of the person, System 4 engenders selfless and wholesale humanitarianism.

Ultimate Purpose

Beyond the worth of an individual life, System 4 longs to know the meaning of life itself. In System 4's mind there must be a "why" for our presence on this planet. Yet, this stance has cosmic ramifications. It implies that there is some meaning, some purpose behind nature. System 4 not only acknowledges that conviction, it builds an entire moral code around it. Our duty, according to System 4, is to grasp the purpose beneath existence, then devote ourselves to its advancement. Evil, in essence, is anything which impedes that purpose. As an extension of this principle, System 4 sees certain things as inherently right, others inherently wrong. Due to this emphasis on right and wrong, sin and guilt loom large in System 4 thinking, and judicial metaphors figure prominently in its language.

Because it perceives an ultimate purpose behind human existence, System 4 ushers in a new outlook on time. It envisions time as "going somewhere," following some preestablished course. System 4 commonly posits an invisible hand behind human events, guiding them in a predetermined direction. The notion that time moves toward an inevitable destiny eventually fosters the idea of "progress," a concept that first appeared in Europe early in the modern era.

Taking Time Seriously

System 4 also inaugurates the writing of history, which supplants the penchant for myth in System 2 and the love of heroic epic in System 3. Neither System 2 nor System 3 possesses true historical consciousness, for unlike System 4, they see no transcendent purpose for existence. If time serves no purpose, no unifying theme, there is nothing to learn from history. In Systems 2 and 3, events of the past have no meaning. They are only curiosity pieces that can be distorted into myth or enlarged into epic without causing them undue harm.

Contrast that to System 4, which wants to know precisely what happened in the past, as free of distortion as possible. Israel, one of the earliest nations to record its history, only embarked on that enterprise once Moses introduced System 4 codes to the tribes. Historians like Herodotus and Thucidydes did not appear in Greece until System 4 ideals began to shape public discourse.

System 4 takes time seriously not only in terms of history, but in hourly affairs, too. System 4 punches clocks and runs things in rhythms keyed to organizational schedules. It no longer thinks of time in seasonal terms, but as a sequence of moments, each one fraught with potential. Because it believes an ultimate purpose directs the universe, System 4 feels a compulsion to manage time wisely in the service of that design. For System 4 "wasting time" is a mortal sin, punctuality a cardinal virtue.

Religion Gets Organized

Religion makes vital strides in System 4. As this modality becomes prominent in a society, fully developed theology emerges for the first time. Spiritual leaders devote immense energy to systematic doctrinal development. Religion begins to center on those who expound the tenets of the faith and are skilled at explaining nuances of the moral code. Rabbinic schools and theological seminaries suddenly spring up. Priestly functions slip into subordinate roles as preaching and proclamation come into their own. (As the Old Testament progresses from System 3 to System 4 themes, for instance, it mentions fewer and fewer high priests, but recounts the work of prophets in detail.) Sects and denominations proliferate, each defining itself along doctrinal lines. The term "heresy" enters the human vocabulary for the first time.

At the cultural level System 4 religion typically struggles long and hard before it secures a triumph over System 3 outlooks. Moses, as we noted, brought System 4 religion to Israel. But centuries later great prophets like Amos and Isaiah still labored diligently to replace System 3 hedonism with a System 4 ethos. Even David, the man after God's own heart, vacillated continually between System 3 and System 4 perspectives. We see this especially in the psychology of the Psalms. At one moment he can appeal to the highest principles and transcendent purposes, then in the next turn back to the warlike, revenge-seeking motives of System 3.

Viewed from outside, nothing is more striking about System 4 religion than its institutional expression. The words "synagogue" and "church" develop such elasticity that they can describe either the building where people worship or the collective body of people who worship there. The phrase "organized religion" becomes almost redundant, for few people can envision religion as anything but organized.

Institutional loyalty, a commonplace throughout System 4, is especially pronounced in religion. In none of the previous systems did people build their personal identity around a particular form of religious expression. In System 4, by contrast, worshipers characteristically think of themselves as Baptists or Catholics, Shiites or Sunnis, Orthodox or Reformed.

The concept of God also takes on new dimensions. System 2 worships the "deity-behind-nature" and System 3 the "deity-who-defeats-the-enemy." System 4 worships "deity-who-is-lawgiver-and-judge." While System 2 seeks to appease God through sacrifice, System 4 seeks to please Him through obedience. It comes as no surprise, therefore, that scribes (or their equivalent) establish themselves as key figures in System 4. As specialists at interpreting spiritual texts, their role is to draw out the implications of divine law so that obedience can be complete. This elevation of law also makes System 4 the first modality in which religion contends with the problem of legalism.

Spiritual Development in Adolescence

Conversions in the evangelical sense of the word occur in System 4. The reason young people often become Christians during adolescence is that System 4 fully activates at roughly the same moment as the onset of puberty. The biochemistry of the two events is probably connected. With this change, the mind can finally manage the realm of the highly abstract. The world of ideas suddenly unfolds. Teens embark on their first political arguments. They develop strong opinions about what is fair and what is not. Concepts like patriotism and social responsibility become comprehensible. They begin to see the "why" behind rules they could not previously understand. Having worked out their sense of self in System 3, they now work out their system of beliefs in System 4.

This means that early adolescence is a pivotal period for spiritual and moral formation. Those years establish the ethical framework that is likely to guide adult life. System 4 builds on systematic thinking and logical deductions. It stresses the importance of knowing one's convictions and standing by them. In a word, System 4 works toward discriminating judgment. Failure to nurture this system during formative years may leave an individual unable to think critically about issues that later systems are sure to raise.

Stability and Authority

While System 2 tribes worry about lapsing into the *personal* chaos of System 1, for System 4 the worrisome threat is falling back into the *social* chaos of System 3. System 4 distrusts System 3's penchant for hedonism and its demand for immediate gratification. System 4 views those traits with disdain, as marks of immaturity. System 4 also raises its eyebrow at System 3's ostentatious display of personal wealth and finery. System 4 calls for simple lifestyles and an uncluttered existence. It wants to distance itself from anything that looks garish or wasteful. Christians who are System 4 dominant want to be known as careful stewards of the blessings God has given. Therefore, they will commonly forego frills in the way they design and decorate church facilities. They will also prefer conservative salary packages for ministers and church employees.

As a reaction against System 3's hedonistic excess, System 4 celebrates the art of postponed gratification. System 4 prides itself on the ability to wait patiently for long-term payoffs. System 3 lives off the spoils, System 4 off its investments. System 4 creates retirement plans and social security systems to provide long-term well-being. Such pursuits never enter the mind of System 3, whose ethic is "eat, drink, and be merry, for tomorrow we may die."

Because of its concern that System 3 will reassert itself, System 4 is highly susceptible to fear. It equates well-being with social stability and is afraid of anything that might upset that balance. To be certain that nothing gets out of hand, System 4 likes lots of checks and balances. It wants everything organized, under control. To that end System 4 is an avid builder of organizations and institutions. It then entrusts authority to those with managerial prowess. System 4 tries to win by out-organizing the competition, not out-muscling it. In politics, System 4 replaces warrior-kings with presidents, prime ministers, and chancellors. In commerce, System 4 brings in the age of the business owner who holds all the reigns of power.

Regulation and Bureaucracy

Because it cherishes stability so highly, System 4 creates a mammoth outpouring of regulations. Their purpose is to keep things orderly. Everyone is expected to know the rules and to stay within them. Whereas System 3 equates losing with shame (and thus does whatever it takes to win), the shame in System 4 is to play unfairly. "It matters not whether you win or lose," System 4 says, "but how you play the game." Such a notion is completely beyond System 3. Nor can System 3 identify with System 4's feelings of disgrace if it wins by cheating.

In every aspect of existence, therefore, System 4 does its utmost to develop guidelines for fair play. Within System 4 political systems, regulatory agencies mushroom. If System 3 builds empires, System 4 follows with bureaucracies. The purpose of this bureaucracy is to clarify policy, promulgate the rules, and police those affected. Whereas System 2 resists change by saying, "We never did it that way before," System 4 is quick to add, "And it goes against policy."

Striving for Predictability

Another factor in System 4's massive regulatory thrust is its drive to create homogeneity and conformity. System 4 tends to believe that stability is fragile unless everyone thinks and acts alike. System 4 is eager to identify the "true believers" and is suspect of anyone not in its camp. It operates on the presupposition that anyone who is not for us is against us. Devoted partisans in the Cold War demanded that every nation take a stand on "the issues." Staunch anti-communists denied that a nation could be non-aligned. Any effort to be neutral, they argued, was tacit support for Marxism. In that regard System 4 is not entirely free from the "us-versus-them" mentality of System 3. System 4 has simply transformed System 3's physical and economic struggle into an intellectual and moral one.

Because it strives for predictability, System 4 is uncomfortable with issues that are gray. It wants everything to be black or white. It should not surprise us, therefore, that judicial professions rise to prominence in societies where System 4 is dominant. No one has the right to "take the law into his own hands" — which is precisely what System 3 always does. In System 4 everyone is entitled to his day in court. We are to settle our grievances before the bar of justice. Should we lose before that tribunal, we are expected to take defeat graciously. Even if we think justice was compromised, we are to value the stability of the system over our desire to settle the score.

To promote that outlook, System 4 fashions its heart and soul around attributes such as loyalty, responsibility, duty, respect for authority, harmony, and trustworthiness. System 4 is highly sacrificial, the first modality in which people will die for a principle or a noble cause. System 4 creates the type of employee every business wants, workers who pride themselves on dependability and integrity, who often "go the extra mile" and do so ungrudgingly.

Positional Authority

Authority in System 4 is a top-down process. As in System 3, those at the top make the rules; those at the bottom follow them. In System 3, followers obey out of fear. In System 4 they obey out of duty. (If they do not, however, System 4 will quickly revert to fear tactics to bring them in line.) Anything less than compliance with authority is tantamount to disloyalty. System 4 may disagree with authority, but will rarely defy it. System 4 reserves defiance only for those moments when its most cherished and fundamental tenets are violated.

Authority, moreover, is positional in nature. We obey the person in charge, even if we question his judgment. The boss is after all the boss. He may know less about the business than half his employees. But by virtue of his position, he has the right to make the decisions. The CEO is "entitled" (note the emphasis on his "title") to set direction, whether he is competent to do so or not. In the parlance of the military, no matter what one thinks of the general's decision, a good soldier salutes and follows orders.

As we shall see in the next chapter, Systems 5 and 6 view authority in an altogether different light. That difference leads to the oft-voiced complaint, "People today just don't respect authority." Examined carefully, such statements usually point to disregard for "positional" authority. System 5 is more concerned with the competency of authority than the title it wears. System 6 evaluates authority in terms of its genuine compassion, not its place on the organization chart. As Systems 5 and 6 view things, authority that is not respected (i.e., does not have genuine credibility) is not true authority in the first place.

Paternalism

Even though System 4 promotes the worth of the individual, it is by no means fully egalitarian. It is a very class-conscious system. One is born to a certain station in life and expected to keep it. As in System 3, everyone is to know his place. It is a disgrace for members of the aristocracy or upper class to marry "beneath their estate." At lower levels of the social ladder there is equal disgrace in marrying someone from a racial or ethnic group that is disdained.

Even System 4 democracy observes class lines. Everyone may have worth, but not necessarily the right to vote. In the System 4 democracies of the ancient Greek city-states only about ten percent of the population was enfranchised. System 4 democracy in the early American republic excluded slaves and women from the ballot box. System 4 enforces the notion of individual rights, but questions the fitness of everyone to rule.

Although this resembles the kind of power elitism we saw in System 3, it is much less exploitive and more prone to be paternalistic. In System 4, to be born to privilege is to inherit the duty of providing and caring for the less fortunate. System 4 thus endows great benevolent institutions, hospitals, and social welfare programs. It builds orphanages, asylums, and sanitariums. But it sees its role as providing resources, not personal involvement. It may therefore give its money without ever taking the interest to see first hand what its money is doing.

Modern colonialism is another form of this paternalism. As System 4 gained ascendancy in Europe, the military empires of System 3 gave way to colonial empires. Many who defended colonialism argued from a paternalistic perspective. Advanced nations, they said, have a duty to protect and guide those unprepared for self-rule. One of the better known expressions of this sentiment was the notion of "the white man's burden," which was a popular concept in the nineteenth and early twentieth centuries.

System 4's Downside

Throughout most of our history as a nation, System 4 has been dominant in public life. System 3 conquered the continent and System 4 organized it. System 4 fought the Depression, the Nazis, the Cold War. It created a cultural mindset in which the solution to every problem was to "throw an organization at it." Government became an alphabet soup of agency names and acronyms. Evangelical and conservative churches also thrived in this period, for their basic outlook coincided with the spirit of the age. Their emphasis on biblical authority touched a responsive chord, as did their call for self-discipline and temperate lives.

But those "glory days" of System 4 also had their downside. When System 4 dominates the scene, it can put too much emphasis on conformity and homogeneity. It tends to suffocate those with an independent or innovative spirit. System 4 may be interested in truth, but it has far more passion for "old truth" than "new truth." It can also become so buried in rules, regulations, and proper procedures that it loses its compassion. We all know stories about justice gone awry when bureaucratic indifference or regulatory nit-picking imposed terrible inequities.

Tender spirits also perceive System 4 as terribly harsh, for it turns to fear and guilt as principal motivators. System 4 manages people with a big stick. It believes in swift and sure punishment for wrongdoing. Only the fear of punishment, it suspects, prevents people from returning to the self-indulgent and unprincipled lifestyle of System 3. System 4 is not so far removed from System 3 to forget that the latter can be deeply recalcitrant. Until it feels the moderating influence of System 4, System 3 yields only to superior force. Mindful of that fact, System 4 uses a club to keep System 3 instincts at bay.

Sermons in System 4 play repeatedly on themes of eternal punishment, fear of divine retribution, the enormity of sin, and the reality of guilt. In a similar vein, System 4 management operates on the motto, "Do that one more time and you're fired." System 4 is far more prone to point out what a worker or volunteer does wrong than offer commendation for things done right. After all, "doing it right" is merely one's duty. And for System 4, duty is a routine expectation. Why praise someone for doing what is only routine?

If System 4 does not learn genuine compassion and encouragement, its message of fear and guilt proves dispiriting, even crushing. Delicate consciences, already staggered by self-condemnation, are its frequent victim. System 4 often saddles them with more guilt than they can bear. Frail egos may also languish when System 4 begins to criticize. Dogged by self-doubt and a gnawing sense of failure, they desperately need reassurance, not a dressing down. Under the weight of System 4 criticism they may lose all motivation to keep trying.

Flight from System 4

As with any regime, when System 4 becomes oppressive, its victims start looking for an escape. Before long they discover that others are also ready to join them in their flight. Not everyone in this exodus feels oppressed by System 4. Many are simply frustrated with it. They are tired of bumping against its rules, regulations, and bureaucracy which seem to thwart all change. System 4, after all, never throws caution to the wind. It seems to change at a snail's pace, if at all. Inevitably, that lethargy becomes more than some souls can stomach. The only way to get things done, they decide, is to bolt the System 4 structure altogether.

Thus, both those who feel crushed by System 4's expectations and those who feel suffocated by its rules move out in search of something different. The former have had their fill of put-downs. They are looking for someone to tell them that they can succeed, then coach them in how to do it. The second group wants a place where innovation and imagination are rewarded, not discouraged. Both will find what they are looking for at System 5, the next stop on their journey.

8

System 5: The Quest for Achievement

The first four modalities have been active across the globe for thousands of years. That has given us ample opportunity to observe them in extremely different cultures. We have watched System 3 build empires both ancient and modern. We have seen how System 4 presents itself in the Moslem Middle East, in Christian Europe, and in Marxist China. With a high degree of confidence, therefore, we can isolate the systems elements of a society from the features that are purely cultural.

When we come to the later systems, that distinction becomes more difficult. We do not have the luxury of studying Systems 5, 6, 7, and 8 across a broad cultural front. So far they have only established themselves in reasonably advanced, technically sophisticated settings. In most cases the cultural context has either been Western or one in which a developing society was consciously imitating Western ways.

It is somewhat challenging, therefore, to peel back the cultural overlay and see the pure essence of these systems. Our descriptions of the remaining modalities reflect that limitation. Forced to use illustrations from contemporary life and politics, our examples will seem to suggest that these modalities only emerge in Western style states. Is that indeed the case? Can they arise under other circumstances? The jury is still out on that issue. We only know that they have appeared thus far in a somewhat limited cultural garb.

The Emergence of System 5

System 5 emerged from transitions that occurred in early modern Europe. Clare Graves dated its first appearance to the fourteenth and fifteenth centuries. Alongside the decline of System 3, those centuries saw the "reurbanization" of Europe following the Black Plague, the first stirrings of the Industrial Revolution, and the birth of modern scientific method. Soon Western governments were fostering a rising tide of capitalism and engineered political structures to support it.

From those initiatives System 5 eventually flowered, putting down deep roots in North America and Western Europe. For several hundred years, however, System 5 predominated fundamentally within the intelligentsia, that tiny minority who were fortunate enough to gain collegiate educations or to be trained in scientific thought. At the level of the masses it took several more generations for System 5 truly to establish itself.

By the twentieth century, however, Americans across the board were ready to forego the alliance of Systems 3 and 4 (by which they subdued the continent) in exchange for a coalition of Systems 4 and 5, by which they would one day go to the moon. In the late nineteenth century John D. Rockefeller, Andrew Carnegie, Leland Stanford, Thomas Edison, and other entrepreneurial giants were the vanguard of System 5 ascendancy. But millions quickly joined their ranks. Within two generations System 5 swept across the entire landscape of American values.

Stephen Covey, author of *The Seven Habits of Highly Effective People*, pinpointed that watershed at the First World War. In his doctoral dissertation, Covey surveyed the success literature that had appeared in the United States since 1776. He found that prior to 1920 most writing on this subject focused on character, "things like integrity, humility, fidelity, temperance, courage, justice, patience, industry, simplicity, modesty, and the Golden Rule," as Covey words it.[1] Notice in this language all the System 4 priorities. Between the First and Second World Wars, however, Covey noted that the literature became focused on the "function of personality,

---

[1] Stephen R. Covey, *Seven Habits of Highly Effective People: Restoring the Character Ethic* (New York: Simon and Schuster, 1989), p. 18.

of public image, of attitudes and behaviors, skills and techniques, that lubricate the process of human interaction."[2] This marked the escalating influence of System 5 perspectives.

Seeking Effectiveness and Efficiency

As a reaction to System 4's caution, System 5 promotes the drive to achieve, to get things moving. It has little patience with policies and structures that hamper what it perceives as needed change. It looks for elbow room, a place to try its wings. It equates security with personal effectiveness. Likewise it throws out any class barriers that would preclude it from rising as high as its own potential merits.

We are deeply indebted to System 5 for the high standard of living in America. It is a tremendously energetic system and extremely inventive. It innovates tirelessly and delights in experimentation. Whatever its investment — whether in time, energy, or money — it wants decisive bottom line results. It therefore puts a premium on freedom, creativity, and efficiency. System 5 believes that anything worth doing should be done with excellence and professionalism. It therefore endlessly seeks ways to simplify processes, streamline operations, and maximize returns.

Education and Scientific Method

Education takes a decided turn toward diversity when System 5 moves on stage. Learning under System 4 dominance, whether seen in Talmudic studies, the Scholasticism of the early medieval universities, or in canon law, was fundamentally authority-based. One learned, not by experimentation, but by systematically exploring what all the great authorities of the past had said on a subject, weighing their respective arguments, and offering one's own conclusion. This was true whether pursuing theology, law, or

---

[2] Covey, *Seven Habits*, p. 19. For an analysis of Covey's thought in this regard, see "What's Effective about Stephen Covey?" *Fortune* (December 12, 1994), pp. 116-126. Note, however, that this article misreads Covey and pegs the watershed at the Second World War, not the First.

medicine, the only three degree fields in the thirteenth-century curriculum.

Once System 5 took over the university, it spawned hundreds of degree offerings. System 5 threw open the door of human learning by setting aside the authoritarianism of the past and ushering in what we today call the scientific method. Instead of exploring truth by comparing the opinions of great authorities, System 5 goes at the task of truth by examining the results of carefully executed experiments. Anything open to experimentation is therefore open to study. All truth, no matter how long held and revered, is subject to re-examination and verification.

Counselors and Consultants

Unlike System 4, which exalts organizational loyalty, System 5 values institutions only to the extent that they empower personal opportunities and fulfillment. In fact, the thrust for personal fulfillment runs all the way through System 5. System 5 devours self-help books. It turned self-help seminars into a multi-billion dollar industry, just as it spawned another multi-billion dollar market for psychotherapy.

Only when System 5 becomes widespread, indeed, is counseling financially viable as a profession. System 4, for the most part, shuns counselors. When the going gets rough, System 4 resorts to its high sense of duty and sacrifice. It tries to lick adversity through perseverance. System 4 loves phrases like "buckle down," "grit your teeth," or "suck it up and keep going."

System 5, on the other hand, seeks to overcome adversity with insight. It surrounds itself with knowledge brokers who provide instant assessments and understanding. The consulting industry flourishes in System 5. So do specialists in every field of endeavor. By promoting specialization, System 5 assures itself that expertise is constantly and immediately accessible.

System 5 also thrives on data. It sees data as the key to fresh perspectives and fruitful insights. Charts and visuals are everywhere in a System 5 society, especially graphics that portray information in a thumbnail fashion so that its significance is quickly grasped. In that sense *USA Today*, with its splashy color, simple-todigest articles, and graphs in abundance, is the epitome of System 5 publishing.

Is There a Better Way?

System 5 is likewise the realm of entrepreneurs who re-engineer entire businesses or industries. System 5 continually asks, "Is there a way this could work better?" It is inquisitive, eager to explore new approaches. That is why it excels at inventiveness. It is fascinated by technology and loves to surround itself with the latest gadgets. It thrives on the cutting edge, at the point where technologies converge in unprecedented ways. Its technical prowess allows it to harness previously untapped sources of natural power, probe mineral deposits deep within the earth's crust, and pull vast harvests from the sea. Thus, the exploitation of nature, which we first saw in System 3, goes forward apace in System 5.

System 5 creates its own brand of heroes. Heroic figures in System 3 perform feats of strength. In System 4 the hero is a giant of intellect, self-discipline, and duty. System 5 draws its heroes from men and women who succeed through shrewd innovation or by "beating the system at its own game." System 5 loves the story of high school dropouts (a symbol of being a misfit in System 4 structures) who go on to build vast commercial empires. For System 4 the path from rags to riches is self-discipline and hard work. System 5 opts for a different path. When it sets its eyes on wealth, it learns how to market itself and leverage its strength.

Social Revolution

Because it prizes innovation so highly, System 5 creates many a technology that induces social and economic revolution. However, System 5's revolutionary thrust is quieter and more subtle than either the military revolutions of System 3 or the political revolutions of System 4. In contrast to those two modalities, which start revolts on purpose, System 5 seldom does. If anything, it is an "unintentional revolutionary." Its revolutions grow out of tinkering with new ways of doing things, little knowing what the outcome might be. Nor does System 5 fret in advance about the social implications of its initiatives.

Take Thomas Edison's invention of the light bulb, for instance. Never once in his experiments did Edison ask if his plans were socially responsible. He did not turn to an assistant and say, "What would it mean to the fabric of American life if grocery stores operated around the clock and people played baseball at night?" He simply set to work, curious to see if he could make a filament glow inside a glass-encased vacuum. The rest, as we say, is history. Few developments have had greater impact on the way we live and structure activities than his little glowing filament.

That is the way with System 5 revolutions. They begin with a technical breakthrough that proves immensely beneficial, especially with regard to comfort, convenience, or time-savings. Seeing those benefits, people eagerly change lifestyles to take advantage of them. One by one, therefore, System 5 innovations transform the way society functions. As a result, unlike Systems 3 and 4 revolutions, which are sometimes overthrown, System 5 revolutions are all but irreversible.

Standing Out from the Crowd

We need not be inventors or innovative entrepreneurs, however, to operate in System 5. The lives of ordinary families show frequent System 5 influence. The techno-miracles that fill our dens and family rooms testify to the System 5 blood in our veins. So do the designer name outfits in the closet. Any time we buy a product merely because it is "new and improved," we are evidencing further System 5 behavior. Yuppies have a quintessential System 5 outlook, as do most suburban neighborhoods.

The marketing industry is a master at targeting System 5. Advertising plays the recurrent theme, "You deserve it. Do it for yourself." Radio and television commercials promise us success, or at least its trappings, if we only purchase a particular product. Advertising also defines success in a decidedly System 5 fashion. To succeed is to free yourself from constraints that others have not escaped. To borrow the refrain from an old Frank Sinatra tune, System 5 likes to say, "I Did It My Way."

The all-American game of "keeping up with the Joneses" is

another indicator of System 5 values. System 5 is particularly susceptible to status-seeking — the right address, the right car, the right college for the kids. Thumbing its nose at homogeneity (a cornerstone of System 4 values), System 5 wants to stand out from the crowd, to separate itself from the herd. The surest way to show its distance from the pack, it believes, is to surround itself with tokens of success. System 5 can therefore spawn crass materialism. In its rawest form System 5 relishes big houses, Rolex watches, and BMWs in the driveway. It pushes its children away from the liberal arts toward professional and business degrees that promise status and financial reward.

The Bigger, The Better

System 5 is an avid resume builder, for it derives its sense of self-worth from its pattern of achievement. Its instincts tell it that "bigger is better," whether in businesses, churches, or bank accounts. One reason it tries to quantify everything is to determine who is on top. "The one who dies with the most toys wins," System 5 jokes. But for many who are System 5 dominant, that line seems more nearly a motto than a humorous aside.

As with System 3, System 5 is highly competitive and keenly aware of winners and losers. System 3 believes competition weeds out the weak. System 5 believes competition makes everyone stronger. System 3 competes for power, System 5 for material success. In both instances winners adorn their life with trappings of luxury and pleasure. System 3 does so to show off its glory, System 5 to confirm its status.

Lots of Options

Second only to effectiveness, System 5 worships variety. It wants countless choices on the shelf. If you are selling to System 5, you must offer options aplenty, whether your product is cars, stereos, college degrees, or midweek Bible classes. System 5 wants more options than it can possibly use. It puts up a satellite dish to snatch a hundred channels from the sky. Then it buys a VCR so it can rent videos on weekends.

System 5 sees options as freedom to choose, to tailor-make

things to personal taste and preference. Even with something as mundane as ice cream, it looks for a menu with 31 flavors. Thus, in contrast to System 3 (which builds empires), System 5 puts up supermarkets and shopping malls, monuments to one-stop variety.

Because of this insatiable appetite for variety, System 5 becomes quickly bored once things quit moving. It tunes out long lectures, especially those with no visuals to punctuate them. It wants business presentations that are succinct and lively. In entertainment it loves action movies, in sports a high-scoring offense. One reason television has wholesale influence on System 5 is TV's skill at adroit pacing. To hold System 5's attention, media programming typically changes camera angles every few seconds to give an illusion of movement, even when the action is static.

Short-Term Commitments

Organizationally System 5 looks for leaders who think strategically. There is, however, an intriguing paradox in this modality. On one hand it appreciates strategic, long-range planning (a trait that acquires even greater importance in System 7). But it has enough kinship with System 3 that it looks for near-term payoffs.

System 5 therefore zigs and zags, making frequent midcourse corrections. It is highly opportunistic, sometimes out of sheer self-interest, at other times because it believes that opportunities to succeed are fleeting and must be quickly seized. System 5 is thus guarded in making long-term promises, either to individuals or to organizations. If an effort does not seem to be working, System 5 is quick to jettison it and look for something with more potential.

To System 4, with its towering respect for long-term constancy, System 5 appears fickle and uncommitted. Indeed, that criticism carries a germ of truth. Marriages fail more often in System 5 than in System 4. System 5 also lacks the strong sense of community that System 4 carefully fosters. System 5 feels a primary duty to its personal potential, only a secondary duty to the community. Or to put it more kindly, System 5 believes that the community will be strongest when everyone in it has achieved genuine self-fulfillment.

Because it does not have the community-building priority of

System 4, System 5 relationships have a way of becoming superficial. Unless it checks its individualism, System 5 may find that it has countless associates, but few close friends. Churches where System 5 is dominant are often friendly, but a difficult place to find friendship. That is because System 5 builds networks, not friendships.

Nor is it particularly adept at (or perhaps inclined toward) "closeness" with others. It joins organizations primarily for two reasons: to expand its network of contacts and to gain new "how-to" skills. When an organization no longer serves those purposes, System 5 commonly drops out. Because it turns to organizations for empowerment, System 5 expects leaders to be teachers, not authority figures. It speaks admiringly of executives who are "mentors" and pays respect only to leaders who enable the success of others.

New Types of Bureaucracies

When it puts its hand to politics, System 5 refashions the entire concept of government. It gives everyone the right to vote. It seats women in Congress, on judicial benches, in the governor's office. It creates populist movements, promotes "sunset laws," and empowers the electorate with tools such as referendum and initiative. In place of the representative government in a System 4 democracy, System 5 democracy demands an accountable government.

System 5 inherits a political structure that conforms to System 4's concept of politics. For System 4 the primary role of government is to maintain national defense, public order, and safety. To those ends System 4 creates a sprawling bureaucracy of regulatory agencies. When System 5 comes along, it adds other vast bureaucracies, but with altogether different aims in mind. The duty of government, System 5 believes, is to serve as a catalyst for the private sector, to enlarge economic and personal potential across the citizenry.

System 5 bureaucracies are therefore promotional, not regulatory. They dispense subsidies, grants, and Federal loans. They underwrite research, then place the results in the public domain for entrepreneurs to exploit. System 5 launched this concept of government in the U.S. over a century ago, when it persuaded Congress to construct a transcontinental railroad. The success of that venture became a compelling argument to use public funds for System 5 initiatives. Today no one thinks it strange when the government builds space shuttles, atom smashers, or superhighways for data.

Personal Spirituality

Religion in System 5 takes on a distinctly individualistic flair. Rather than centering on a weekly cycle of corporate worship, System 5 spirituality focuses on "my personal walk with the Lord." Publishers find a burgeoning market in System 5 circles for works that tie biblical principles to marriage, child-rearing, and effective daily living. Books on spiritual disciplines and personal devotion also sell well. System 5 relishes the parables of Jesus about rewards for exceptional service, not to mention the exhortations to personal excellence found in the Epistles.

System 5 does not have the interest in doctrinal theology that we saw in System 4. Instead, it promotes practical theology. It values sermons largely to the degree that they translate into clear, workable applications. System 5 seeks out Bible classes that are "relevant." It feels no obligation to be part of an experience simply because that event appears on the church calendar. In fact, System 5 spirituality may function quite well without the direct benefit of congregational membership. System 5 draws much of its edification from mentors whom it follows via tapes, Christian radio, books, and periodicals. It may find those learning experiences so profitable that it sees little point in regular church involvement.

All of these influences work to weaken organizational loyalties in System 5 religion. Church hopping is frustratingly characteristic of worshipers who are System 5 dominant. They may visit a church consistently for months, but never consider placing membership. Asked why, they will answer that they are hesitant to make long-term commitments. System 5 wants to keep its options open.

Where System 5 prevails, people also move freely from denomination to denomination, from one spiritual heritage to another. Commonly they do so several times in adulthood. They can cross these lines with ease because judgmentalism relaxes in the transition from System 4 to System 5, and with that relaxation, doctrinal rigidity recedes.

Nor is System 5's personal identity bound up in the organization it belongs to, as is the case with System 4. System 5 is less concerned with finding "the church that is right on the issues" and more concerned with finding "the church that is right for me." Essentially this means a church that helps me become what God calls me to be. If System 4 emphasizes believing God's words, System 5 emphasizes developing His gifts. System 4 works out the implications of truth, System 5 the implication of talents. For that reason churches with a System 5 texture endeavor seriously to involve every member in vital service and ministry.

The Coming of the Megachurch

One of the most significant contributions of System 5 religion has been the emergence of the church growth movement. Its fascination with data, demographics, trends, ways to quantify success, and effective methodology are all extensions of System 5 thinking. Related to that, System 5 churches have placed a premium on ministers who not only teach well, but who manage well. A well-managed church, they believe, will be a growing church.

Again as a result of System 5 influence the church growth movement has encouraged churches to market themselves professionally. System 4 churches tend to look at that development askance, as a resort to manipulation, not legitimate evangelism. System 5 merely sees it as being more creative and effective in competing for the hearts of people.

True to its "bigger is better" mindset, System 5 has underwritten the emergence of megachurches. Congregations of five, ten, or fifteen thousand are now found in most metropolitan areas. Interestingly, the vast majority of these are unaligned with mainstream denominations, which are, after all, bulwarks of System 4 institutionalism. Megachurches tend to have names that do not identify them with any particular spiritual tradition. They become a mecca for those who are seeking alternatives to the church in which they were reared.

Again in keeping with System 5 values, megachurches offer a host of options. They provide multiple worship services, often with differing styles for each one. They promote dozens or even hundreds of study opportunities, not just on Sunday morning, but throughout the week. Their buildings convey a definite sense of status and success. They pride themselves on professionalism throughout their organization. In a word, megachurches are the flowering of System 5 religious culture.

Churches need not be large, however, to show System 5 markings. As System 5 gains influence, congregations move to more prestigious addresses, often under the stated guise of becoming more visible and accessible to the community. They begin to build their staff around ministerial specialists — youth ministers, education directors, family counselors. They often set a higher standard of education for their pulpit. Fellowship activities take on new importance because "people just don't know one another the way they used to." And the church vies with organizations in the community for the time of its leaders. This problem does not restrict itself to leaders, however. Throughout the congregation, scheduling becomes a far greater challenge, because everyone is so busy.

System 5 Pitfalls

For churches both big and small the materialism of System 5 can prove seductive. The so-called "health and wealth gospel" is simply System 5 thinking with a spiritual topping. Churches where System 5 is dominant often find it easier to raise money for buildings than for missions. They may be more interested in funding counseling for their members than evangelism of the lost. They want Christianity, like everything else in their life, convenience-packaged so that it does not intrude too greatly on their schedules.

Despite its high level of achievement, therefore, System 5 holds great potential for harm. Left unchecked, its materialistic bent subsumes the spiritual. Moreover, to achieve its ambitions

System 5 is prone to use people, then discard them. In this regard System 5 can be as ruthless as System 3. Companies and churches that operate dominantly in System 5 are notorious burnout factories. Businesses driven by System 5 demand long hours and sacrificed holidays of their employees. But when these incessant demands lead to emotional or physical breakdown, the worker may be quickly abandoned.

At a personal level, System 5 is often blind to damage it does to key relationships. Because it tends to be self-preoccupied, System 5 must be careful not to alienate close associates and family, or even lose them altogether. Discovering all too late the difference between a full schedule and a full life, System 5 spawns many a midlife crisis. Those who have lived by the System 5 credo learn that no amount of status equates with feeling worthwhile, that immediate gratification is no substitute for ultimate satisfaction. A gnawing hollowness sets in, the sense of lost fulfillment that jazz singer Peggy Lee once captured in her soulful tune *Is That All There Is?* Or take Tony Bennett's song, *The Good Life*, which in a single line captures both the upside and the downside of System 5. The good life means "being free to explore the unknown, like the heartaches, when you find you must face them alone."

Sometimes System 5 devotees discover this emptiness through their own bitter experience, sometimes through the mistakes of their peers. In either event, if they conclude that they have paid too dear a price for System 5 status and success, they may decide to call the deal off. Surrounded by relationships in shambles, they long to repair or undo the damage. They try to find settings where they can start anew, building a shared life with others. They enlist themselves in causes that seek to prevent the very abuses they once perpetrated themselves. With that commitment, however, System 5 is being left behind as System 6 begins to beckon.

9

System 6: The Quest for Intimacy

Among other things, System 6 is a response to the elitism which System 3 introduces and which subsequent systems prolong. System 3 builds on power elites, System 4 on class elites, and System 5 on success elites. Each creates its own brand of victims. In System 3 the worship of power leads to exploited masses, having no voice in their destiny and often subjected to harsh conditions. The enforcement of rules and regulations in System 4 can crush sensitive spirits by slipping into heartless legalism, bureaucratic insensitivity, and rigid authoritarianism. And the quest for prestige and status in System 5 exaggerates the gulf between the "haves" and the "have-nots."

When System 6 comes on the scene, it takes up the cause of these victims. System 6 sees itself as a healing presence in a deeply injured world. Its bumper stickers read, "Envision World Peace" and "Have You Hugged Your Child Today?" In contrast to System 5, which seeks self-empowerment, System 6 works to empower the entire community by undercutting the elitist remnants of previous systems. It works to replace alienation and indifference with unity and wholeness. System 6 defines well-being as a genuinely caring community where everyone bonds together in a lifestyle that is ecologically and socially sensitive.

Egalitarianism and Bonding

System 6 is therefore notably egalitarian. It builds coalitions and consensus, not power pyramids. It takes up the banner of those who are powerless or disenfranchised. From its ranks come the foot soldiers for all the "rights" movements — civil rights, women's rights, animal rights, etc. System 6 goes overboard to be certain that everyone has both a place and a voice at the table. System 6 is eager for dialogue and listens for feelings as much as viewpoints. Its goal is to identify sensitivities and protect them from roughshod treatment.

System 6 also seeks a sense of "oneness" with others. It is trying to shed the highly individualistic lifestyle of System 5, that frequently degenerates into selfishness. System 5 typically has lots of associates, but relatively few close friends. When its world of status and achievement proves unfulfilling, it becomes gnawingly aware of its loneliness and detachment from people. As System 5 transitions into System 6, therefore, bonding intimately with others becomes a compelling drive.

With its egalitarianism, System 6 is also attractive to those who got the short end of the stick in the previous modalities. Here they are promised an opportunity to be heard, to belong, to have as much voice in their destiny as anyone else. As opposed to the world of System 5, where everyone must compete on his or her own merits, where the fortunate get ahead and the rest are left behind, System 6 promises membership in a supportive, cooperative team. When the group succeeds, everyone in it succeeds.

Changing the Workplace

In business this outlook manifests itself in the concept of self-directed work teams and efforts to create so-called "horizontal corporations." Because it views multi-tiered organizations as too bureaucratic and impersonal, System 6 always pushes to flatten the management hierarchy. Since System 6 sees the success of the individual as dependent on the success of the group, it tries to involve everyone in the workforce in setting goals and objectives for the enterprise. It also tries to give each person a voice in the decisions that affect his or her work team.

Affirmation, Concern, and Support

System 6 excels at creating activities that other systems describe as "touchy-feely." It promotes sensitivity training and is a front-line advocate of multiculturalism and political correctness. System 6 insists on non-judgmental acceptance and an atmosphere of mutual encouragement. Thus, where System 5 reads self-help books, System 6 forms self-help groups. Twelve-step programs are its offspring, as are therapy groups for rape and incest recovery, (although the twelve-step approach, with its emphasis on affirming a Higher Power, has strong System 4 tones).

System 6 organizes blood drives and volunteers for the Red Cross. It forms food banks, builds shelters for the homeless, and joins Habitat for Humanity. System 6 puts together medical mission campaigns for churches and free dental clinics for indigents. It goes door to door in the neighborhood soliciting for the March of Dimes. When natural catastrophe strikes, System 6 turns out en masse. It basks in the bonding of lives that occurs as people rally together in the wake of disaster. To be sure, one hazard in System 6 is vulnerability to care-giver burnout.

System 6's compassion for victims extends to the planet itself. System 6 finally revolts against the exploitation of nature that System 3 begins and System 5 carries to unprecedented heights. In an effort to reverse damage to the Earth, System 6 presses for recycled materials and renewable resources. It cleans up oil spills and protests landfills. It passes legislation to protect endangered species. It demands environmental impact studies before any project can go forward. System 5, indeed, views System 6 as anti-business, for System 6 places ecological concerns above profitability.

Looking for Intimacy

Hollywood, long enamored of System 6, is its avid promoter. Entertainment figures vie with one another to endorse System 6 causes. Movies and television eagerly expound System 6 themes. Directors routinely "humanize" characters by dressing them in System 6 garb. In a recurring theme that film-makers love, high-flying System 5 executives (usually male and mean-spirited) walk away from career and financial success when System 6 begins to throb in their breasts. Normally the story line tosses in a little midlife crisis for good measure.

This way of depicting transitions from System 5 to System 6 is decidedly stereotypical. But the stereotype punctuates just how different the priorities are between System 5 and System 6. As opposed to System 5, System 6 puts little emphasis on economic security, professional achievement, or personal recognition. What System 6 wants is inner contentment and peace. To find that harmony of the soul, System 6 believes, we must rediscover our true self in a life shared intimately with others.

To speak of intimacy, however, is to invite misunderstanding. Americans use the phrase "being intimate" as a euphemism for love making, an equation which suggests that sex creates intimacy. Intimacy in System 6, on the other hand, may be totally unrelated to sexuality. It may also have nothing to do with marriage. Rather, it is the blending of lives in a setting of mutual and supportive sharing, an opening of the inner self to others so that we build an experiential bond with them. To the degree that sex is part of such a relationship, it is an outflow of intimacy, not the creator of it.

System 6 Benevolence

This one-anotherness makes System 6 particularly appealing to spiritually sensitive hearts. Almost every contemporary congregation has significant pockets of System 6 in it. System 6 starts small group ministries. It opens subsidized day care centers. It develops recreational centers for the elderly. It takes youth groups into neighborhoods to paint homes for impoverished widows. Every year System 6 argues for more money in church budgets to benefit the poor and needy.

Of course, System 6 is not the first system to be benevolent. System 4 and System 5 also work to relieve want. But System 6 stands apart from them in terms of motivation and tirelessness. System 4 benevolence flows from a paternalistic sense of duty and responsibility. ("I have been richly blessed. I therefore have an obligation to share my bounty with those less fortunate.") System 5 is generous for a number of reasons, some having to do with genuine gratitude, some with assuaging guilt for a selfish lifestyle. System 5 will also use generosity to enhance its own prestige. ("I want people to see my name associated with this good work.") System 6, by contrast, tries to relieve need out of pure compassion. And because it cares so keenly, System 6 continues to labor long after System 4 and System 5 have gone home.

Experiential Spirituality

In the realm of religious expression, System 6 is highly experiential. Worship, it believes, should touch deep emotions and enlarge the feeling of community. System 6 warms to the spiritual impact of music and drama, but downplays preaching, especially the variety that delves into doctrinal intricacies. Exegetical nuances strike System 6 as centered too much on the head and not enough on the heart. System 6 accepts the possibility that such teaching has its place, but not as a centerpiece of public worship. It is no surprise, therefore, that System 6 bears poorly with dogmatism and denominational line-drawing. System 6 likewise wants nothing to do with a spirit of exclusivism.

Because it needs a feeling of community, System 6 spirituality seeks out interpersonal settings. It wants to be involved in genuine fellowship. But System 6 does not equate fellowship with being a member of a particular congregation. System 6 is looking for relationships, not an organization to join. Too much organization, indeed, makes System 6 uneasy.

As a consequence, System 6 has no more institutional loyalty than System 5. System 6 may change churches often as it searches for the "right" spiritual atmosphere. As a minimum, System 6 wants an environment that respects diversity and where messages (both verbal and non-verbal) are uplifting, affirming, and unifying. System 6 walks into a church expecting true warmth and acceptance. It watches closely to see if relationships are loving and supportive. It also takes note of a church's care for the "4-H club" — those who are hungry, hurting, handicapped, and helpless.

System 6 Vulnerabilities

In a word, System 6 is a well-meaning system. It wants to make the world a healthy, humane place. It believes its efforts blaze the way toward a kinder, more compassionate planet. Unfortunately, that is not always the outcome. Instead of overcoming alienation, System 6 may actually aggravate it. System 6 has a penchant for estranging the other modalities. In part this is because System 6 is inclined toward a "more caring than thou" attitude. (Or in the case of System 6 churches, a "more spiritual than thou" attitude.) When System 6 looks at how Systems 3, 4, and 5 have influenced history, it sees an unbroken trail of victims. "Such insensitivity!" it exclaims. But if System 6 fixes on that thought, it starts feeling morally superior to previous systems. Once developed, that air of superiority leads to condescension, if not contempt. The disdain may be so subtle that System 6 is oblivious to it. But the other modalities sense it immediately. They also have a name for it — arrogance.

Internal Contradictions

Critics of System 6 also toss other pejoratives its way, words like "narrow," "rigid," and "intolerant." Blinded by its own self-perception, System 6 typically dismisses such charges out of hand. How could anyone think it guilty of the very behavior it abhors? But as we shall see, a certain element of truth underlies these indictments. There is often glaring incongruity between System 6's promise and its reality.

This inconsistency arises from structural problems in System 6 itself. To be specific, System 6 ideology and System 6 methodology are not always compatible in the same context. They get in each other's way. That compels System 6 to make trade-offs, either compromising its ideology on one hand or its methodology on the other. Whatever the choice, System 6 ideals inevitably suffer. To illustrate, we can look at one of System 6's most difficult balancing acts, managing its simultaneous commitment both to diversity and to consensus-building.

Wanting to distance itself from the elitism of previous systems, System 6 calls for everyone to have a voice in decisions. It is this egalitarianism that gives impetus to consensus-seeking in System 6. It hesitates to move forward until everyone feels good about the direction the group is taking. Now, so long as goals remain narrowly focused and viewpoints somewhat aligned, this stratagem works reasonably well. Once diversity exceeds certain bounds, however, consensus-building becomes cumbersome, if not impossible. As issues multiply and viewpoints diverge, time constraints preclude a voice for everyone. (I know the dilemma well. My family faces it every Sunday as we pass umpteen restaurants while trying to reach consensus on where to eat.)

Limiting Diversity

System 6 must therefore make a choice. It can either violate its own egalitarian values by creating a decision-making hierarchy and imposing a decision. Or it can set narrowly defined goals, then admit into its enterprise only those with views similar to its own. In other words, it can limit diversity. Given those choices, System 6 usually opts for the latter. Having decried elitism, it now begins to exclude those whose views it cannot assimilate. Before long it does not offer a ready hearing — or even a particularly friendly one — to anyone who questions its conventional wisdom. Having espoused diversity, System 6 ends up managing it poorly. At that point onlookers find it difficult to differentiate between the elitism that System 6 disdains and the exclusivism that it practices.

This aspect of System 6 behavior is what fuels the charge of narrowness and intolerance. System 6 seems to post a sign above its entry: "Restricted area. Access granted to System Sixers only." Even being a fellow Sixer is no guarantee of admission. System 6 organizations can evolve into such closed circles that outsiders, Sixers included, cannot break in. As strong affinities develop among those already inside the circle, the entire group tends to become inwardly focused. Members enjoy each other so much that they simply fail to notice newcomers. Visitors to System 6 churches often observe, "They seemed so loving among themselves. But no one spoke to me." The result is a church that thinks of itself as warm and loving, while others view it as cold and self-centered.

Choppy Political Waters

The narrow focus of System 6 organizations also impairs effectiveness in the political arena. When trying to forge a far-reaching political alliance, System 6 often is unable to work with other groups, including those committed to System 6 causes. Early in his presidency Bill Clinton enjoyed a massive System 6 following. His running mate, Al Gore, openly embraced the System 6 agenda of the so-called "green movement." Yet, within a year of the Clinton inaugural news magazines were reporting green in-fighting around the Vice President. Some green organizations were even scolding him for compromises he had made to get their own initiatives through Congress. *Newsweek* quoted one White House official, himself a longtime advocate of System 6 causes, as lamenting the attacks from his former colleagues in environmentalism. He compared them to "a football team with only one play: a 95-yard touchdown pass."[1] As Gore learned, the politics of half-a-cake is not part of the System 6 repertoire.

This is ironic, for System 6 prides itself on being an exemplary collaborator. Yet it does not have a good track record at engineering broad-based coalitions that endure. In politics System 6 is most effective when it builds strong moral suasion around a specific issue, then hammers through narrow reforms. And "hammer" is the appropriate word. Because it dislikes "striking a deal," System 6 must often become shrill and combative to win its reforms. System 6 may try to wrap itself in a banner of peace and harmony. But in the give-and-take of politics it can also show a vicious streak.

Moreover, when System 6 tries to take the moral high road, its critics swarm from the woodwork. System 4 is likely to lead the swarm. System 4, with its strong sense of absolutes, views System 6 morality as selective and arbitrary. System 6 insistence on non-judgmental acceptance (at least within its own circle) strikes System 4 as ethical relativism. And often the charge is well founded. Under the guise of values clarification, System 6 is known to undo System 4's studied effort at values indoctrination. Further, to maintain consensus, System 6 may simply avoid matters that could potentially divide its ranks. Anyone wanting to

---

[1] "Barbarians Inside the Gate," *Newsweek*, Nov. 1, 1993, p. 32.

save the whales is welcome to join the group, so long as they talk about nothing but saving the whales. Again, to System 4 this "ducking of issues" smacks of ethical indifference.

Pluses and Minuses

Like all other modalities, therefore, System 6 has its strengths and its weaknesses. The cause of humanitarianism is clearly indebted to System 6, as is the effort to save our ecology. Throughout His ministry Jesus endorsed many values which System 6 seeks to elevate. His concern for the downtrodden, His insistence on compassion, and His identification with outcasts are a microcosm of System 6 ideals.

On the other hand, System 6 eventually disillusions many of its own kind. Independent spirits who depart too far from the "group-think" of System 6 may find themselves on the outs with their fellow System Sixers. Others, wanting to see wholesale change, become put out with System 6 ineptness in the political arena. Then there are those with a sense of impending crises who become frustrated with System 6's slow and inefficient consensus-building.

These disillusioned fellow-travelers do not write off the in-justice and injury that are the object of System 6 concern. Nor do they minimize the need to avert ecological disaster. But they do launch out in search of other approaches to those problems. Once they had hoped System 6 would end polarization. Instead, System 6 exacerbated it. Once they had believed it would heal all wounds. Now they feel it has betrayed them. Clearly, there is healing to be done. But there must be a way to do it in a more holistic, less polarizing manner than System 6 has found. The search for that better way leads eventually to System 7.

Systems 7 and 8:
The Quest for Holistic Solutions

With this chapter we step onto the threshold of the future. As we examine Systems 7 and 8, we catch a glimpse of the way people will likely think in the early twenty-first century. Today only about one person in eight is predominantly a System 7 or System 8 thinker. A hundred years ago almost no one was. But the influence of these systems is rising rapidly. They will probably become dominant more quickly than any of their predecessors. Global communication and trade will have much to do with that. So will the relentless exchange of ideas which today's technology makes inevitable. But primarily they will gain ascendancy because they, more than any other systems, are willing to tackle the full complexity of today's world.

System 7

Standardized testing indicates that only about ten percent of the population is System 7 dominant.[1] Whether you are part of that number or not, one thing is certain. You are clearly open to System 7 concepts. Otherwise, you would have put this book aside long ago. From the opening chapter we have been looking at the church through System 7 eyes. To the degree you have followed our analysis and found it helpful, something of System 7 is already in your blood.

---

[1] Based on norms established by the National Values Center in Denton, Texas.

System 7 emerges once the pace of change becomes so accelerated that it is often easier to see chaos than order in events. Tom Peters captured the essence of System 7 in the title of his book *Thriving on Chaos*.[2] System 7 presumes that nothing is nailed down, that everything is in a state of flux. Other modalities may think they can roll back the tide of change. System 7 never does. It expects change to continue and to accelerate, making unprecedented demands on human adaptability. System 7 believes that we must be endlessly flexible. Otherwise change will overwhelm us.

To System 7's way of thinking the key is knowing where and when to adapt. To that end System 7 continually looks for patterns that are not immediately apparent, but may indicate what lies ahead. It seeks those patterns from two vantage points. First is a high angle perspective that allows System 7 a big-picture view of the world. System 7 examines life through a wide-angle lens. For that reason, generalists frequently fare better in System 7 than specialists. System 7 looks for connections and relationships everywhere, even between widely separated events. It knows that causal elements are not always found in close proximity (whether measured in time or space) to the outcomes they influence.

Complex Causality

System 7 also peers deep beneath the surface, to detect the flow and counterflow of underlying forces. As a result it develops models that are often more intricate than those that evolve from other modes of thinking. That intricacy reflects the very environment that gives birth to System 7. By the time System 7 activates, whether within a society or in the life of an individual, the immense complexity of life is pressing down upon us. We are recognizing that problems are far more complicated than they once appeared.

Because of that realization, System 7 is no longer satisfied with simple cause-and-effect explanations. Instead, it looks for multiple lines of causality. Its elaborate models are an extension of that outlook. In the same vein, System 7 also seeks multifaceted solutions. System 7 is suspect of "simple fixes," believing them

---

[2] Tom Peters, *Thriving on Chaos: Handbook for Management Revolution* (New York: HarperCollins, 1987).

doomed to fail. It perceives problems as an immense puzzle, not with a single piece missing, but with a host of pieces that need to be put in place simultaneously.

The notion of causality itself undergoes change in System 7. In quantum physics (a System 7 view of nature), probability statements replace Newtonian concepts of cause and effect. No longer do we describe sub-atomic particles as being at a specific location. Instead we speak of the probability that they will be found within a certain range of locations. Again, things are not "nailed down." That is one reason System 7 models are more intricate than those from earlier modalities.

Promoting Systems Harmony

Because System 7 sees wholesale survival at stake, it believes that we can no longer afford for various modalities to war with each other. Everywhere it turns, System 7 sees important institutions at risk, in no small part due to inter-system conflict. One reason you are reading this book, no doubt, is concern about escalating unrest in the church. At a global level System 7 looks out on a polarized world in which humanity, armed to the teeth, stands poised to kill itself. Alarmed at that prospect, System 7 does its utmost to avoid a headlong rush to disaster. This survival-mindedness is so reminiscent of System 1 that Clare Graves suggested that System 7 is revisiting the existence issues of System 1, but at a societal, not a personal level.

From System 7's perspective the key to survival is finding a way to manage the viewpoint of all the systems so that they interact as an integrated whole. Rather than working *against* each other, which has been their historic practice, they must learn to work *with* each other. As far as System 7 is concerned, the risks are too high to consider any other alternative. System 7 therefore senses a compelling need to develop new organizational concepts that hold the promise of system harmony.

Holistic Approaches

Consistent with that desire for harmony, System 7 tries to go lightly through the earth. Its ideal is to be as unobtrusive as possi-

ble, especially with nature. System 7 is likely to use organic fertiliz-
ers in its garden and to choose biological counter-measures to
combat pests. System 7 promotes solar and wind power. It fights
for greenbelts in cities. It works on development codes that
require construction to blend into the surrounding terrain.

As an integrator, System 7 is the first holistic system. It has an
almost intuitive appreciation for the six previous systems. It
believes that each of them has a rightful role in a healthy, well-
balanced society. System 7 not only affirms them, however. It
enjoys interacting with them (if they are healthy). This sets it apart
from earlier systems, which often eye each other with disdain. Not
so with System 7. If an occasion calls for System 2, System 7
throws its arms around System 2 and exults in the warmth of the
moment. If the situation calls for System 3, System 7 can knock
heads with the best of them. Yet System 7 is equally at home
standing alongside System 6 as an advocate for victims of violence.

Organizational Flexibility

In its commitment to flexibility, System 7 has no rigid organizational preference. This, too, sets it apart from earlier
systems. System 7 may create a hierarchical pyramid one day, an
egalitarian task force the next. It looks at the people and the job at
hand, then determines what approach is most appropriate. System
7 never considers itself locked into an operational mode, even one
it toiled hard to create. When it senses a change of circumstances,
System 7 restructures immediately.

Unlike System 5 businesses, with their massive inventories and
sprawling warehouses, System 7 business operates on just-in-time
delivery systems. In many cases System 7 products do not exist
until the customer asks for them, then they are generated immediately. This type of business thrives on instant delivery of highly
customized products, like music stores that now allow customers
to create their own cassette tapes and compact discs, on the spot,
by combining selections from any number of albums and artists.[3]

---

[3] For a very readable description of System 7 business theory, see
Stanley M. Davis, *Future Perfect* (New York: Addison-Wesley Publishing Co., Inc.,
1987).

To maintain this type of quick-response capability, System 7 wants as few structural restrictions as possible. It designs organizations that can be quickly assembled, quickly modified, and quickly dismantled. "Virtual corporations" are a striking example of this strategy. Core staffing in a virtual corporation may consist of only a handful of people. But they network themselves with dozens of other specialized groups. No formal links tie these groups together until a customer wants a specific service. At that point the core staff taps elements throughout the network to form a working alliance. Once the project is complete, the alliance immediately dissolves itself.

System 7 is perfectly comfortable with this type of scenario. Other modalities are not. System 4 needs structure to feel secure, and the virtual corporation offers little of that. Nor does this kind of organization provide the obligatory corporate ladder for System 5 to climb. System 6 has its own misgivings about the virtual corporation, for it sees no possibility for community or consensus in this kind of put-together, tear-apart company.

For most of us, then, it is quite a stretch to be completely comfortable in a System 7 world. Fortunately, as with all the systems, we can benefit from System 7 concepts even if we are not yet fully at ease with all of its qualities. The virtual corporation may not be your cup of tea, but you can still become a proficient practitioner of System 7 techniques, especially systems thinking.

People who excel in System 7 typically have broad, diverse interests touching on numerous (and often unrelated) fields. They are perpetual learners. They are also self-confident and self-determined. They appreciate the recognition of their peers, but function well without it. They listen to adverse criticism, but do not brood over it. System 7 is typically difficult to intimidate, for it is relatively free of fear. Failure does not frighten it, nor does an uncertain future or the prospect of wholesale change. System 7 is confident that it can learn quickly enough and adapt readily enough to manage whatever comes.

Information, Systems, and Processes

If anything worries System 7, it is the possibility of responding to change too slowly. After all, survival is at stake. Like an over-the-horizon radar, System 7 continually probes the distance, eager to see what may be lurking just beyond our vision. Trend analysis is its favorite pastime. It never tires of books by futurists. System 7 is a tireless consumer of information. It wants access — immediate access — to vast informational resources. The more, the better. System 7 operates on the premise that knowledge precedes under-standing, and access precedes knowledge. System 7 engineers the Electronic Village. It signs up for CompuServe and Prodigy. It logs onto Internet. It swaps tidbits on electronic bulletin boards. Then there is the so-called information superhighway, promoted by System 5, which sees a fortune to be made from it. But the dream belongs to System 7.

Perhaps it is no coincidence that System 7 has gained as-cendancy at the same time computer networks are becoming commonplace. We cannot describe computers without referring to systems and processors. System 7 uses a similar process-centered vocabulary. Because of its focus on the flow of processes, System 7 creates mental models that are highly dynamic. They stress interaction and change. This distinguishes them from more common types of models (like blueprints and organization charts) that show static or spatial relationships, not interactive ones.

To attain vital goals, System 7 tries to bring careful quality control to all critical processes. System 7 has given us key concepts in the Total Quality Movement, especially in the realm of process improvement techniques. And because quality enhancement is a long-range endeavor, System 7 businesses are more willing to sacrifice near-term profits for long-term profitability than are their System 5 counterparts.

Oversights and Shortcomings

Ironically, the very thing that gives System 7 its strength is also a potential pitfall. Because System 7 thinks so much in "big pictures," it is not particularly astute at managing details. Without intending to do so, it simply overlooks them. Neither is it noted for its people skills. To get an unbiased view of what is happening, System 7 is given to detaching itself from the immediate moment.

It tries to rise above the fray and park its feelings and emotions. But that detachment works to distance System 7 from engagement with people. System 7 can become so preoccupied with processes and interactions that it neglects the most important interaction of all — person to person.

Despite its vaunted flexibility, moreover, System 7 has difficulties of its own in dealing with other modalities. For one thing, it becomes impatient when partisan spirits in the other modalities revert to parochialism. System 7 offers its suggestions, but may promptly walk away if it gets a hostile or indifferent hearing. System 7 is searching for kindred spirits who want to get things turned around. It may see no point in wasting time with those whose primary interest is defending self-interest or turf.

In addition, System 7 is so pragmatic that other systems (System 4 in particular) often think it unprincipled. That may or may not be the case. The apostle Paul, who hardly lacked for principles, saw the value of System 7 strategies. "I became all things to all men," he wrote, "so that I might win them for Christ." That sounds very much like System 7's willingness to adopt System 5 approaches one moment, a System 4 style the next, dependent on the dictate of circumstances.

Spiritual Outlooks

Paul's guidance on church organization also shows a band of System 7 thought. By keeping organizational mandates to a minimum, the apostle liberated church structure from a particular cultural moment. In effect he freed church leaders to work within prevailing modalities. Where Systems 3 and 4 are dominant, churches usually expect authoritarian leadership. System 5 and 6 congregations, on the other hand, decry authoritarianism. They want more egalitarian leaders. The genius of the New Testament is that we can have it both ways, without violating biblical directives.

When it comes to religion, System 7 looks for groups who are holistic in their approach. Spiritual formation in System 7 is also holistic. System 7 is as likely to seek religious inspiration in art, poetry, and literature as in formal works of theology. Because

System 7 can show genuine respect and regard for those who espouse values or ideas completely contrary to its own, it expects a spiritual climate where contrasting views receive a cordial hearing. System 7 reads in religious traditions far removed from its personal heritage, looking for commonalities between its own faith and what it finds in the faith expression of others.

This eclectic spirit in System 7, however, has not stripped it entirely of provincialism. System 7 grapples with survival issues in a somewhat localized context. It is concerned first and foremost with the survival of the institutions, social system, and nationalistic entities that are dearest to its heart. Yet, we are starting to see people emerge from System 7 and bring a genuinely global perspective to the survival impulse. Those individuals (currently only a fraction of one percent of the population) seem to be the trailblazers as the mind prepares to accommodate System 8.

System 8

We will make only a handful of observations about System 8, primarily because it has only limited impact on today's church. In the decades ahead that impact will grow as System 8 itself becomes more common. To give you a foretaste of things to come, therefore, we want to devote a few paragraphs to the direction System 8 seems to take us.

If you want to think of one term that epitomizes the System 8 perspective, you might opt for the recently popularized phrase "global village." System 8 looks at the entire planet as a single living unit. Political boundaries and ethnic enclaves disappear from sight for System 8. In its eyes all humanity forms a single unit, struggling with one ultimate issue, namely, finding a means by which we can all survive on a planet where resources are finite, the ecological system fragile, and human populations mushrooming. System 5 may speak of the planet as Spaceship Earth, but System 8 thinks of it as Life Raft Earth. All mankind is crowded onto this fragile craft, making its way across the unforgiving oceans of space.

An Olympic Village World

In many ways this sounds like System 8 is merely an enlargement of System 7, which is itself deeply concerned about the threat of disaster. But there are important differences. System 7 focuses on dangers to society, its institutions, and specific ecological structures. These dangers, as System 7 sees it, are the product of polarization and narrow perspectives within the various modalities. System 8 has a far broader concern. It sees the ecosystem itself in jeopardy, primarily because the entire human community is acting nearsightedly.

System 7, especially in its earlier phases, seeks solutions primarily within traditional political and institutional structures. System 8 is convinced that we must move beyond those structures to find ultimate solutions. Thus, in its effort to build a sense of shared destiny that unites all mankind in one spirit, System 8 looks for ways to circumvent ideological and political barriers.

To illustrate, we might think of the exceptional spirit that flows through the Olympic village as nations of the world come together for the winter or summer games. During the 1994 Winter Olympics, commentators often reminded their audiences of Sarajevo, the idyllic site of the winter competition just ten years before. But now Sarajevo's great athletic facilities had been reduced to ruins, victims of incessant shelling. Looters, desperate for combustibles to keep their families warm, had furthered the destruction. System 8 is driven to find some way to forestall a global Sarajevo. It believes the communal and unifying spirit of the Olympic village must be enlarged to encompass the entire earth and maintained on a permanent basis.

Theology Re-Enters the Dialogue

Many of the leading figures in System 8 are men and women at the very frontier of theoretical physics and astronomy. In terms of public name recognition, Stephen Hawking (who first advanced the theory of black holes in space) is probably the best-known of these. But you will find System 8 outlooks in fields far removed from physics. Managerial theorists are promoting ideas that originated in the worldview of System 8. The last section of Peter

Senge's *The Fifth Discipline* encourages managers and executives to open themselves to viewpoints with distinct System 8 overtones.[4]

Theology is also showing the first indications of System 8 influence, for this modality holds the promise of reintroducing spiritual dialogue to public discourse. System 8 science operates at the point where dividing lines between thought and matter become essentially indistinguishable. It therefore invites a complete re-examination of metaphysics, a discipline that all but disappeared in the twentieth century. Now, with an impetus from System 8, this discipline is staging a return. But ever since the dawn of Greek philosophy, metaphysics has been a breeding ground for theology. Today, interestingly enough, theorists in the field of cosmology are openly talking about something akin to a God-principle for the first time in ages.[5]

To say the least, this reintroduction of God-talk has been highly controversial. In the late 1970s a national leader in astrophysics made a widely debated speech in which he described efforts to unravel the mysteries of the distant past, to revisit the dawning moments of the universe. We have pressed back into the distant recesses of time, he said, until now we seem to be scaling the last slope, beyond which we will be able to see what happened at the very instant the cosmos began. And as we approach that final summit, he added, we seem to hear the songs of theologians coming from the other side of the ridge.

For months following those remarks angry letters appeared in scientific journals around the world, denouncing any incursion of theology into the realm of science. But in the years since, other leading voices have spoken out about the spiritual essence that they perceive as an underlying reality. If this trend continues (and forces in System 8 seem to indicate it will), the next century may prove to be one of the most creative spiritual periods in human history.

---

[4] Peter M. Senge, *The Fifth Discipline: The Art & Practice of the Learning Organization* (New York: Doubleday, 1990), pp. 368-71.

[5] We are seeing books, for instance, that even put God-language in their titles, which was quite unthinkable just a few years ago. One such title is Leon Lederman, *The God Particle: If the Universe is the Answer, What is the Question?* (New York: Houghton Mifflin, 1993). See also Freeman J. Dyson, *Infinite in All Directions* (New York: Harper & Row Publishers, 1988).

11

A Systems Summary

This chapter encapsulates the key features of each system in a highly summarized fashion. When used in conjuction with the systems overviews in chapter three, these summaries offer a one-stop, quick-reference guide to all the modalities. There is little in this chapter by way of new material. You can therefore bypass this section, if you wish, without missing essential information. On the other hand, taking a few minutes to review these descriptions is an excellent way to reinforce the key concepts we have already covered.

SYSTEM 1

Primary Existence Issue: Physical survival in the face of immediate threats to my very life.

Organizational Impulse: Random groupings of people in bands that forage together for food, water, and shelter.

Leadership Structure: Virtually non-existent. Things are done merely as short-term reactions to external events.

Family Expression: Focused largely on finding basic necessities. Everyone, down to young children, pitches in to lay hands on food.

Spiritual Expression: Little or none. When people reenter System 1 in times of wholesale disaster or protracted battles with disease, prayers are survival-centered.

Learning Style: Conditioned response to events in the environment. Otherwise, no real learning experiences.

Characteristic Activities: Wanders aimlessly. Responds primarily to the drive of appetites. Accepts conditions others would consider degrading in order to survive.

Responds Warmly To: Settings that hold the promise of a steady supply of food and shelter. A full stomach. A warm bed. People who sense its plight and come to its rescue.

Responds Adversely To: Things that require it to think beyond the next few hours. Judgmentalism about its condition.

Strengths: A very powerful drive that enables survival in almost unthinkable conditions.

Weaknesses: Lives in the world of immediate necessities. Contributes nothing to the physical well-being of society. Hordes scarce commodities. Has no sense of principle or duty to others.

SYSTEM 2

Primary Existence Issue: Personal safety in a world of unseen powers.

Organizational Impulse: Forms family or tribelike groups that share a common sacred or "safe" place.

Leadership Structure: Little hierarchy. A circle of elders governs the life of the community and maintains its traditions. A chieftan (or patriarch of the clan) either doubles as a "priest" for the group (like Job sacrificing for his children) or has someone with great "spiritual prowess" at his side.

Family Expression: Thinks of the clan as family, so that cousins are almost like brothers, nephews almost like one's own son. Observes distinct rites of passage. Limits marriage options rigidly to avoid mixing of bloodlines. Defends family honor passionately and is quick to avenge a wrong done to the family.

Spiritual Expression: Centers on the immanence of God, i.e., a God who is near at hand in the ordinary events of life. Senses the wonders of God in the mysteries of nature. Has a profound respect for things sacred. Flourishes in an atmosphere of rich ceremony, ritual, and symbolism. Builds a close sense of communion with Deity, but also fears God's wrath.

Greatly concerned about being cursed by God. Thinks of sin in terms of defilement. Casts salvation in terms of cleansing and purification.

Learning Style: A passive learner who looks to "parent" figures as models of appropriate conduct and behavior. Needs repetition, rituals, and routines to learn effectively.

Characteristic Activities: Spiritual rituals. Tribe-building ceremonies and celebrations. Imaginative stories. Repeated rhythms in music and dance. Use of totems and charms. Frequent gatherings at a shared "safe place" or "holy place."

Responds Warmly To: Tradition. Sensing the presence and blessing of divine benevolent power. Maintaining routines that keep life predictable.

Responds Adversely To: Individualism. Questioning of received ways. Disregard for spiritual forces. The threat of curse. Sudden change or dislocation.

Strengths: High respect for spiritual realities. Builds a strong sense of "belongingness" with the group and creates vibrant family ties. Appreciates the importance of mystery and awe in personal experience. Is deeply attracted to the wonders of nature. Shows great creativity with symbols.

Weaknesses: Permits little freedom in determining one's role in life. Highly vulnerable to gross superstition. Does not think critcally. Extremely resistant to change. Has often led to strong ethnic rivalries.

SYSTEM 3

Primary Existence Issue: Physical safety in the face of hostile human forces.

Organizational Impulse: Builds strong hierarchies in the shape of a pyramid where privilege and luxury are exclusively for those highest on the pyramid.

Leadership Structure: A tough "boss" at the top with a distinct pecking order of subordinates, each ruling his portion of the domain with a firm hand.

Family Expression: Family fights as a unit against outside threats. Boys raised to be tough, to "take it like a man." Rites of passage built around tests of courage, strength, and endurance. Women treated largely as property, with little voice beyond strictly domestic issues.

Spiritual Expression: Respects God for His power and might. Seeks God as a personal defender in a world of hostile forces. Builds impressive houses of worship, laden with grandeur. Attracted to religious events that are replete with pomp and ornate ceremony. Little given to what we normally think of as theology. Interested instead in how God will act, at this moment, on behalf of His people. Thinks of sin as acting in a manner unworthy of the great God who rules over us. Casts salvation in terms of deliverance from personal enemies (as seen in many of David's psalms).

Learning Style: A manual learner, i.e., learns by handling things, building things, tearing things apart to see how they work. Needs strong, demanding teachers who are not afraid to maintain firm control. Loses motivation without frequent rewards for learning.

Characteristic Activities: Impulsive and pleasure loving. Defiant of convention. Plays to win. Fights frequent turf wars. Exploits weakness. Has little concern about long-term consequences. Loves to show itself daring and fearless.

Responds Warmly To: Being known as tough. Basking in the glory of triumph. Wearing the scars of victory. Showing people up.

Responds Adversely To: Being humiliated (not to be confused with being defeated). Cowardice. People who want a "soft" life. Intellectuals.

Strengths: Provides the fighting might to withstand oppression. Mechanically inventive. Builds strong teams that can react decisively to provocation or threat. Provides a counterbalance to predatory elements in society. Operates with minimal overhead or bureaucracy.

Weaknesses: Easily becomes undisciplined. Does not adequately consider long-range consequences. Can be hedonistic to a fault. Is often exploitive of others. Manipulates people. Often loses sight of mercy and compassion. Creates embittered enemies.

SYSTEM 4

Primary Existence Issue: Moral and social stability in a world given to hedonism, impulse, passion, and violence.

Organizational Impulse: Creates highly vertical organizations, with clear lines of authority from top to bottom. Decision-making is concentrated in authority figures at the top, who promulgate rules and regulations, often through a multi-layered bureacracy.

Leadership Structure: Entrusts authority to leaders who have demonstrated integrity and moral fiber, then follows those leaders almost unquestioningly.

Family Expression: Home provides the focal point for character formation and moral training. Demands strong respect for parents. Sees the wife as a covenant-partner, but not an equal with the husband. Abhors marital unfaithfulness.

Spiritual Expression: Reveres God as the transcendent Author of Truth and Eternal Creator. Profound respect for Scripture and biblical authority. Draws its moral foundation from biblical principles. Produces highly developed theology. Starting with divine mandates, works out implications in great detail. Promotes simplicity and purity of motives in worship. Thinks of sin in judicial metaphors, as guilt for the violation of God's laws and standards. Casts salvation in terms of exoneration and justification.

Learning Style: A passive learner who sits at the feet of authorities and listens. A good aural learner. Enjoys lecture and can follow involved oral presentations. Eager to know what is true, what is false. Diligent about study and homework. Loves the world of books.

Characteristic Activities: Devoted to ideals. Promotes strong moral codes. Goes the extra mile. Works diligently. Keeps covenants faithfully. Insists that one's word be one's bond. Demands truthfulness, integrity, and civilized behavior. Enforces rules of propriety and etiquette. Makes long-term investments and sacrifices. Promotes a vibrant sense of neighborliness. Defends the status quo.

Responds Warmly To: High moral character. The cause of truth and justice. Loyalty. Doing one's duty. Being responsible. Punctuality. Perseverance. Martyrdom for the cause of truth. Stability. Structure. Lifestyles that are unpretentious.

Responds Adversely To: Undisciplined lifestyles. Wastefulness. Dishonesty. Laziness. Disrespect for authority. Excessive pleasure-seeking. Thumbing one's nose at convention.

Strengths: Builds strong communities. Elevates respect for life and human dignity. Puts a priority on truth and learning. Builds extensive moral foundations. Is thoroughly self-sacrificial. Pursues an orderly and well-disciplined lifestyle. Needs few near-term rewards to stay motivated.

Weaknesses: Sees things as black-and-white. Can become oppressively bureaucratic. Slips easily into legalism. Relies excessively on guilt and fear for motivation. Is quite guarded about change.

SYSTEM 5

Primary Existence Issue: Personal success and achievement in a world whose demands for conformity thwart one's inner sense of fulfillment.

Organizational Impulse: Creates competency-based organizations that depend on efficiency and bottomline effectiveness to survive. Mission statements, strategic thinking, corporate goals, and departmental objectives are critical to maintaining organizational focus.

Leadership Structure: Entrusts leadership to people who can make things happen and get strong bottomline results. Leaders are expected to be coaches and mentors, not "bosses" in the traditional sense of the word.

Family Expression: Family is primarily the nuclear family (Mom, Dad, and the kids). Closeness to the extended family is greatly diminished. Wives are equals, typically pursuing professional careers outside the home. High expectations for children to get into the right schools and do well in the right professions. Extreme mobility. Family may rarely eat together, certainly not at breakfast or lunch. Few things done "as a family." Instead, each member has his or her own recreational, avocational, or educational pursuits independent of others in the household.

Spiritual Expression: Looks to God as Friend and Guide. Stresses "my personal walk with the Lord." Concerned with finding opportunities to develop spiritual gifts and use them in the service of God. More interested in practical theology than doctrinal theology. Builds houses of worship that bespeak status and success. Wants professionalism in the way the church goes about its work and the way it presents itself to the community. Thinks of sin as failure to live up to the potential God places within us. Casts salvation in terms of regaining the image of God and being transformed into His likeness.

Learning Style: Wants an instructor who is a mentor and has proven his or her competency. Loves case studies, problem-solving, and testing of ideas. Learns visually, especially from graphics that simplify complex bodies of information. Likes to express views freely during the learning process. Needs a teacher who is an authority, but not authoritarian.

Characteristic Activities: Pursues self-improvement projects. Does things professionally. Is given to symbols of status. Thrives on data. Maintains an energetic schedule. Continually explores new ideas. Enjoys material comforts. Builds profit-driven companies.

Responds Warmly To: Being recognized for achievement. Perks. Having freedom to innovate. Rewards commensurate with performance.

Responds Adversely To: Incompetency. Stifling of innovation. Uninformed leadership. Cumbersome rules and regulations. Make-work activities. Boring presentations. Not being consulted on key decisions. Slow decision-making processes.

Strengths: Highly inventive and innovative. Is thoroughly ambitious and results-oriented. Expands material well-being. Promotes research and experimentation with new methods and technologies. Builds mission-focused activities. Makes an immense enlargement of personal freedoms and options. Gives rise to the capitalist system and to a large middle class in society.

Weaknesses: Easily slips into crass materialism. Often becomes so driven to succeed that it sacrifices key relationships in the process. Can put more emphasis on symbol than substance, more energy into building image than building integrity. Promotes excessive burnout rates. Gets too busy to be neighborly.

SYSTEM 6

Primary Existence Issue: Building bonds of intimacy and mutual support in a world given to insensitivity, alienation, and exploitation.

Organizational Impulse: Fosters small, egalitarian groups that bond intimately together.

Leadership Structure: No hierarchy. Flat organizations. A facilitator leads the group, and individual members may rotate into the facilitator role on an ad hoc basis. Makes decisions, whenever possible, by consensus.

Family Expression: A household of equals. Wives often retain their maiden name. Parents maintain an emotionally open environment that encourages children to talk freely about their feelings. Little or no corporal punishment of children. Family-time activities are anticipated with excitement and carefully protected on the calendar.

Spiritual Expression: Seeks God as a Healer and Reconciler. Wants intimate settings for worship and non-judgmental acceptance of all who gather in that setting. Needs worship that touches deep feelings and causes people to be introspective about their duty toward those who are hurting or powerless. Drawn to the compassion which Jesus had for people who were socially marginalized. Thinks of sin in terms of alienation from God and from one another. Casts salvation in terms of rebuilding relationships both with Heaven and one's fellow man.

Learning Style: An interactive learner who gains new insight by being part of a group in which everyone shares personal experiences and feelings. Learns best in settings of no more than a dozen people or so. Insists that each viewpoint in the group be heard respectfully.

Characteristic Activities: Care for victims of abuse and trauma. Insistence on treating everyone as equals. Disdain for organizational structure. Taking pains to recycle resources. Promotion of self-help groups.

Responds Warmly To: Candor. Demonstrations of genuine compassion. Helping those who are powerless. Coming to the aid of victims. Relieving hunger and want. Promoting ecology and conservationist efforts.

Responds Adversely To: Elitism. Insensitivity to others. Racism. Exploitation. Ostentatiousness. Authoritarianism. Self-aggrandizement. Things that exacerbate the gulf between the "haves" and the "have-nots."

Strengths: Provides a genuine "humanizing" effect on society. Works tirelessly for human well-being. Promotes equality of opportunity across the citizenry. Makes "insiders" of "outsiders." Works toward a healthier planet for future generations.

Weaknesses: Depends on consensus-building, which can be clumsy and slow. Can be so concerned for victims that it allows itself to become gullible. Is often uncompromising in political situations. Tends to disparage tradition, to the point at times of throwing out almost all conventions of society. Can become narrowly focused on one or two issues and shrill in its advocacy of them.

SYSTEM 7

Primary Existence Issue: Averting the looming disaster of a polarized world in which rigid viewpoints and partisan spirits promote warring camps and thwart the flexibility we need to survive.

Organizational Impulse: Creates highly flexible, thoroughly modular organizations that can be restructured and revamped almost instantaneously with minimal loss of momentum.

Leadership Structure: Dispersed decision-making throughout an organization in which information networks tie everyone together in a neural structure.

Family Expression: Household members frequently exchange roles. Family "routines" are continually open to renegotiation and change. Little pressure to conform to some family image. Broad freedoms of choice throughout the household.

Spiritual Expression: Seeks God as the Great Integrator, who brings all things together as a functioning whole. As interested in divine processes as it is in divine principles. Draws spiritual inspiration from many sources not traditionally thought of as "religious literature." Relishes the study of paradox in spiritual truth. Enjoys exploring other religious traditions and identifying common denominators in their faith expression and one's own. Thinks of sin as acting nearsightedly with undue concern for long-term harm to others and to nature. Casts salvation in terms of learning to work harmoniously with the processes God has built into physical, human, and spiritual nature.

Learning Style: Enjoys self-directed learning experiences. Thrives on training that is flexible and permits easy adaptation to individual needs and interests. Uses technology avidly as a tool for learning.

Characteristic Activities: "Big-picture" views. Broad interests. Fascination with information technologies. Long-range forecasting. Open acceptance of people as they are. Unthreatened by change.

Responds Warmly To: Opportunities to be self-directed and self-paced. Studies of the future. Bringing about timely change. Helping diverse elements work together harmoniously. Instant access to vast information sources.

Responds Adversely To: Parochialism. Inflexibility. Narrow viewpoints. Unprincipled decisions. Incompetent leaders. Activities
that leave scars on nature. Simplistic explanations for complex
events. Disinterest in learning and exploring. Being forced to
maintain routine and repetitive patterns over protracted periods of time.

Strengths: Anticipates change well in advance. Works contentedly in
a non-static, ever-changing workplace. Excels at strategic planning. Easily gleans meaning from immense volumes of information. Loves to learn new things. Always open to new alternatives.

Weaknesses: Can be so farsighted that others cannot share its
vision. Often becomes so process-centered that it loses sight of
people and their needs. While looking at the big picture and
broad-based trends can sometimes miss details that call for
immediate action. Does not stick with projects when others fail
to appreciate its counsel.

SYSTEM 8

Primary Existence Issue: Creating a genuine sense that all
humanity is one race, living in a single village, providing equal
access to the planet's resources, but caring for the earth as a
fragile life-partner.

Organizational Impulse: Alliances of highly diverse elements who
lend their expertise to the solution of problems that transcend
ethnic and national lines.

Leadership Structure: Collaboration of leaders who think in holistic terms and who are driven to resolve macro-issues. The leadership function requires pronounced skills in long-range thinking and exceptional abilities to synthesize enormous amounts
of disparate information.

Family Expression: Offspring raised as "children of the world."
Minimal emphasis on ethnic or nationalistic identity. Sacrifices
personal interests to the well-being of all creatures.

Spiritual Expression: Seeks God as the Being behind all being.
Highly metaphysical in its approach. Recaptures the wonder
and mystery about nature that are also integral in System 2.

Wants to build personal "connectedness" with the Well-Spring from which all existence flows. Thinks of sin as failure to treat life in all its forms with due care and respect, a failure which puts us at cross-purposes with the Life Principle behind everything that is. Casts salvation in terms of gaining cosmic awareness of who we are in the greater scheme of reality and acting responsibly as a life-partner with all that exists.

Learning Style: Interactive dialogue with other "macro-issue" thinkers. Immediate access to any information required to address global issues. Needs an atmosphere that encourages people to envision totally unprecedented ways to structure life on the planet.

Characteristic Activities: Trans-cultural and trans-ethnic friendships. Genuine interest in all peoples of the world. Finding ways to circumvent political and organizational barriers that thwart global action. Profound respect for the life-principle of the universe itself.

Responds Warmly To: Integrative solutions. Wholesale sensitivity to the earth's plight in a world of overpopulation and resource depletion. Earth renewal projects. Intuitive breakthroughs that permit sweeping new alternatives to be envisioned.

Responds Adversely To: Insensitivity to the environment. Ethnic or nationalistic self-centeredness. Failure to act with an eye to obligations we have to all humanity. Relegating global needs to subordinate status.

Strengths: Sees the far-reaching impact of actions that others would mistake as having only local import. Keeps political powers aware that threats to human existence indeed loom on the horizon. Is void of the particularized loyalties that have pitted men and nations against one another since time immemorial.

Weaknesses: Thinks so esoterically that many people cannot relate to its insights. Becomes impatient with those who do not share its concern for global survival. Requires vast technological resources to sustain communication flow among widely separated collaborators. Tends toward solutions that are immensely expensive.

12

When Systems Join Forces

Our next objective, now that you have an overview of the eight conceptual systems, is to put that understanding to work. We want to go inside day-to-day congregational life, examine the systems we find there, and explore why they slip into conflict. Before we do that, however, we need to address some preliminary issues. These take the form of four key questions.

- How can I know what my own dominant system is?
- Which modality is best?
- What is the difference between thinking systems and personality types?
- What is the relationship between thinking systems and personal maturity?

Identifying Dominant Modalities

As you worked through the system descriptions, beginning in chapter five, you probably found some of yourself in several modalities. That is because all of us have several systems inside us, not just one. But which of those systems is dominant? Which one has the greatest influence on your decisions and outlooks?

One way to answer that question is through formal testing. There are evaluative tools that we use in seminars and training sessions to help participants identify their personal systems pattern. But you can often accomplish much the same thing on your own. All you need do is observe and listen. Ask yourself questions like these. If I am working on a project and something truly crucial is on the line, what kind of team do I want to be on? How do I want it organized? If I am in charge, how would I run it? If I am not in charge, what role do I want the leader to play? What will it take for my work on this project to be fulfilling?

Listen to your responses. What type of management and organizational style did your answers imply? More than likely that style aligns with one of the conceptual systems. If so, you have an indicator of your dominant modality. Or look at the issue of fulfillment. What is required in the project for you to feel fulfilled?

- Coming out on top of the competition (System 3)?
- Doing your duty well (System 4)?
- Recognition for what you personally achieved (System 5)?
- Developing a fraternal bond with fellow-workers (System 6)?
- Being perceptive enough to anticipate changing conditions and stay one step ahead of them (System 7)?

These, too, are good indicators. Continue your self-inventory by looking at other issues. For instance, which of the following is most likely to be your reason for criticizing a sermon?

- It was too soft (System 3).
- There was not enough Bible in it (System 4).
- It was not practical and helpful enough (System 5).
- There was no feeling and heart in it (System 6).
- The principles were too simplistic and inflexible (System 7).

Or when a problem comes up in your congregation, which of these responses are you most likely to make?

- We need to get a policy worked out on this so everyone knows what to do (System 4).
- This kind of thing happens because our way of doing business around here is so terribly outmoded (System 5).
- We really need to get everyone together who has been upset by this and find a way we can all work together (System 6).
- There are probably several factors that got us into this situation, and we need to take them all into account before we opt for a course of action (System 7).

We could add other examples, but these are sufficient for the moment. They point out the kinds of questions to ask in ferreting out your dominant system.

You might also try this exercise. Our remaining chapters offer concrete suggestions on managing various systems within a church. As you work through those recommendations, monitor your internal reaction to them. When we talk about organizing Bible classes around modality preferences, notice the class that strikes the most resonance within you. When we talk about coordinating volunteers, think about the kind of chairperson you would probably be. Before long your own system map will emerge.

What you may discover is that two or three systems are competing for your loyalty. Perhaps you are torn between System 4 values and those in System 5. Or between System 5 and System 6. This can happen when we are in transition from one modality to another. It also occurs when we have more than one dominant system. If you find that you identify most closely with Systems 4 and 6 and that both influence you equally, then you are probably System 4/System 6 dominant.

Determining the Best Modality

Early in the going, as people hear about conceptual systems for the first time, someone inevitably asks, "Which modality is best?" The response, of course, is, "Best for what?" Over the past few chapters you have seen that each system has its strengths and each its weaknesses. System 3 fights wars better than System 6. System 4 gives itself to sacrificial causes more readily than System 5. System 8 thinks more globally than System 7. Thus, before we can know which system is best, we must identify the immediate and long-range objectives to be served.

At one time or another every modality is "best." That is why a healthy society (or a healthy church, for that matter) needs systems diversity. When the modalities each contribute their unique strengths, life enjoys peak moments. Think about the systems that contribute to a high school football game on homecoming weekend. Thursday evening gets things started with a huge bonfire and pep rally. We are bringing the tribe together, so System 2 is out in force. We reforge the sense of tribal oneness with chants, totems (we call them mascots), and carefully observed traditions like singing the alma mater.

The next evening at 8 o'clock System 3 lines up at midfield. Good friends may play on opposing teams, but right now they are not thinking about friendship. When they go nose to nose in the line, each will have a singular objective — knocking the other on his backside. But System 3 is not alone on the field. System 4 is out there, too. It wears black and white striped shirts, carries a flag in its pocket, and runs around blowing a whistle. It knows the rules of fair play and will enforce them unswervingly. With so much System 3 energy building to a boil, System 4's job is to keep things from getting out of hand.

Needless to say, there is also a definite System 3 atmosphere in the stands. Just listen to what people are yelling! They are urging the young Turks on the field to kill the other team, to stomp the adversary in the ground! We would be hopeful, of course, that these yells are merely symbolic. Surely the fans do not *really* want their sons to kill the adversary. But there are occasions when System 3 sheds its symbolism and becomes a literal presence in the stands. Fights do occur between supporters of opposing teams. So to keep System 3 under control in the bleachers, a few more System 4 elements circulate in that area. Instead of striped shirts, these System 4 cohorts wear badges and carry night sticks. They make their way through the crowd, warning the rowdies and checking to see that no one is smoking or drinking something illegal.

Down toward the end of the field is an ambulance. Inside it are some System 6 specialists. They hope that this will be a dull night for them. But in the event someone is hurt, they are poised to move to the rescue. Even though they are pulling for the home town boys, they will treat injuries on the other team as if the player were one of their own. And what about System 5? Where is it in this scene? Oh, there it is, back behind the bleachers, selling hot dogs for two dollars apiece. System 5 is also nervously pacing the sideline, for a big time college may offer the coach a contract if he produces a winning team this year.

Our football game serves as a microcosm of society. If we had nothing but referees, there would be no thrill of competition. If no one did anything but attend pep rallies, the coach would never have a winning season. On the other hand, a winning season also depends on the esprit de corps from pep rallies and bonfires. Every modality contributes an essential component. When multiple modalities work effectively together, they add new richness and excitement to our existence. And this is just as true for the church as it is for society in general.

So which modality is best? The answer is situational. What are the needs of the moment? What are we trying to achieve? The "best" modality will be the one that serves those purposes most fully. But no system is "best" irrespective of the situation. Nor, for that matter, are any systems "better" than others in any absolute sense. Being better or best is always relative, based on what we seek to accomplish. In a situation where we are trying to innovate, System 4 is better than System 2. But System 5 is better still. On the other hand, if our objective is to create a sense of Holy Presence, System 4 is better than System 5. But System 2 is better than both of them.

No Place for Type-Casting

Just as it is inappropriate to rank the modalities as good, better, and best, it is equally unwise to think of them as personality types. In seminars, when we start to sketch the systems, someone frequently makes the aside, "This is just another way to type people." It is rather natural to slip into this error, for we are conditioned to think in those terms. Over recent decades both management theory and psychology have popularized behavioral models that use some form of type-casting. "Jim is an analyzer." "Jill is an evaluator." "Anna is an introvert." "George is an extrovert."

Although such models may be helpful, they nonetheless convey a "static" sense. They describe something that *is*, not something that is *happening*. By nature, however, a system is dynamic, not static. And that holds whether we are talking about a respiratory system, an air conditioning system, an ecological system, or what have you. The whole purpose of a system, including each of the conceptual systems, is to effect interaction among processes.

Moreover, type-casting always depends on an underlying polarity. To the degree you are an extrovert, you cannot be an introvert. Those are opposite poles. Or again, to the degree you are impulsive you cannot be self-disciplined. The conceptual systems, by comparison, do not define themselves by means of polarities. Just because you have a well-developed System 5 does not preclude you from having an equally developed System 4.

In the strictest sense of the word there are no System 4 people, just as there are no System 5 people. Instead, there are individuals for whom System 4 is currently the most dominant of several active modalities. We lose sight of that dynamic with statements like, "He's a System 6," or "She's a System 5." Those statements come dangerously close to treating the systems as types. And if we ever cross the line into type-casting, we are no longer thinking in systems terms.

On the other hand, if you find it beneficial to use behavioral types, there is no need to abandon that approach. Simply be aware that you will find any of the classic "types" in each of the conceptual systems. Every modality has its extroverts, its introverts, its aggressive personalities, its passive followers. Those traits will not change as a person moves from one dominant system to another. Or to look at it another way, the person who is a self-starter in System 3 will likely be a self-starter in Systems 4, 5, and 6, as well.

Types do prove helpful, we should add, when we move from talking about *intra*personal systems and begin to focus on *inter*personal systems. They give us added information that suggest how a person may respond to specific circumstances and individuals. Our goal, therefore, is not to criticize the value of behavioral typing. We merely want you to distinguish typing from systems thinking.

Modalities and Maturity

If we should not look at the modalities as types, neither should we confuse the later systems with greater maturity. Yet that mistake is common when people discover thinking systems for the first time. What causes this confusion is the order in which the systems activate. Because lower numbered modalities dominate the early years of life, it is easy to associate them with childishness. We might conclude, then, that higher numbered systems evidence maturity.

But what does it mean to speak of someone as "mature"? Maturity admits to a broad range of meanings. Sometimes it refers to nothing more than accrued age, as in the phrase "the mature years." On other occasions it carries the notion of advanced learning or experience. A "mature athlete" is someone who has outgrown rookie mistakes. Then there is maturity that connotes living as a responsible, well-balanced adult.

In none of these senses is there a necessary connection between one's dominant modality and personal maturity. Systems 5, 6, and 7 may activate early in life, late in life, or not at all. Thus, age-maturity does *not* determine their appearance. Nor does gaining greater educational or experiential maturity. High school dropouts can become System 6 thinkers, while a Ph.D. can function in System 4 for life. And as for being mature from the standpoint of responsible adulthood, system dominance again has no bearing on the issue. People can be irresponsible and childish in *any* modality. When talking about modalities, therefore, be cautious with the word "maturity."

Patterns of Development

In a society like the United States, basic social and civic duties (like serving on juries and respecting constitutional processes) require System 4 outlooks, as a minimum. As a result, we expect System 4 to be vibrant by the onset of later adolescence. Elsewhere on the globe, systems development has a different timing. In less complex societies System 2 may be sufficient to meet the needs of most adults. There a person enters adulthood with only the first two systems highly energized. Within that cultural context System 2 may dominate the rest of life.

On the steppes of Mongolia, among the large nomadic clans, System 3 characteristically holds sway. System 3 also prevails in pockets of ethnic warfare across the planet. In settings like those the developmental process may take adolescents only through the first three systems. Nature is preparing them for a social context in which adult roles require little more than System 3 thinking.

This is another reason for not mistaking modalities with marks of maturity. If we define maturity as the ability to manage adult duties within the conventions of a given society, a System 2 thinker can be fully mature in a System 2 culture. Transplanted to a System 4 society, however, he would be ill prepared for adult duties. Social maturity is thus a relative term as we move across the globe.

Enlarged Sophistication

In addition to everything else we have said, we miss a vital reality if we correlate particular systems with maturity. Every system has the potential to become more sophisticated as we grow older. System 3 may be amorally egocentric in pre-adolescence. But in a more adult form it provides the competitive drive to empower our aspirations. System 2 may start out with puerile fantasy. But it also underlies the oneness with God that people experience in advanced levels of faith.

Andrew Greeley made this point quite tellingly in a recent *New York Times Magazine* article.[1] He did not use systems language, to be sure. But we can summarize his argument in straightforward systems terms. He was saying that System 4 religious expression, even when fully developed, does not replace the need for insights only System 2 can give. At the same time, the very act of dialogue with System 4 brings a new sophistication to System 2.

> While institutional authority, doctrinal propositions and ethical norms [note the System 4 values] are components of a religious heritage — and important components — they do not exhaust that heritage. Religion is experience, image and story before it is anything else and after it is everything else.

He then continues:

> If one considers that for much of Christian history the population was illiterate and the clergy semiliterate and that authority was far away, one begins to understand that the heritage for most people most of the time was almost entirely story, ritual, ceremony, and eventually art. So it has been for most of human history. So it is, I suggest (and my data back me up) even today.

---

[1] Andrew Greeley, "Because of the Stories," *The New York Times Magazine*, July 10, 1994, pp. 38-41.

And then, as if he had been looking over our shoulder as we wrote this chapter, he underscores the very dynamic we have been discussing.

> It may seem that I am reducing religion to childishness — to stories and images and rituals and communities. [Listen to the System 2 elements in that sentence.] In fact, it is in the poetic, the metaphorical, the experiential dimension of the personality that religion finds both its origins and its raw power. Because we are reflective creatures we must also reflect on our religious experiences and stories; it is in the (lifelong) interlude of reflection that propositional religion and religious authority [System 4 values] become important, indeed indispensable. But then the religiously mature person returns to the imagery, having criticized it, analyzed it, questioned it [again notice a System 4 methodology], to commit the self once more in sophisticated and reflective maturity to the story.

To use systems terminology again, Greeley is saying that we impoverish religion if we reduce it to the expression of a single modality. For religion to have its deepest impact, multiple systems must be in dialogue with both our experience and each other. To that end, having multiple systems in a congregation is not a detriment, but a boon. Properly engaged, that diversity can spawn fresh energy and creativity that invigorates the entire Body. Moreover, the earlier modalities, through their interaction with the later ones, gain sophistication and breadth they could never achieve alone.

Systems Synergism

We could also look at Greeley's comments using the concept of synergism from systems theory. Synergism is the principle that two systems or processes can produce more when harnessed together than when standing alone. For instance, a twin engine aircraft can carry more cargo than two single-engine planes. Or two researchers, tackling a problem together, usually discover more timely solutions than they would if they worked separately, then merged their results. When one becomes discouraged, the commitment of the other keeps him focused. Should one hit a conceptual roadblock, the other offers an observation that sparks new creativity.

What Greeley points to, then, is a synergistic effect that occurs when we bring System 2 and System 4 together in a religious context. Through their interplay with one another, they produce a type of spiritual expression that we would never achieve by tapping into System 2 religion and System 4 religion separately.

A Multi-Lens View

This type of synergism prevails throughout the conceptual systems, not just with Systems 2 and 4. Figure 12-1 is one way to represent what happens when multiple systems interact. You may recognize this graphic as a slightly modified form of Figure 4-2, which we explored in chapter four. Here we have replaced the matching pairs of arrows in Figure 4-2 with completed ovals. But the ovals still represent individual modalities. In this case we will also think of these ovals as lenses, each ground to a different refraction.

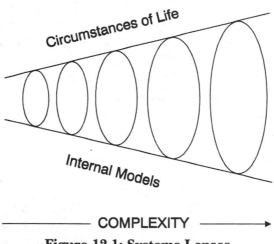

COMPLEXITY ⟶

Figure 12-1: Systems Lenses

Now, imagine yourself taking just one of these lenses and holding it before your eye. The result is a particular view of the world. Next, exchange that lens for another, and your worldview becomes different. But what happens if you look through both lenses at the same time, as we do with a telescope? You suddenly obtain a new, unique view. It differs from what you saw with either lens individually. Yet it also differs from merely adding together the views you got from the individual lenses.

Something similar occurs when we utilize multiple modalities simultaneously. Our mind surveys the world, not through the lens of a single thinking system, but through a combination of lenses. In Figure 12-1, therefore, we can imagine that the various systems are lenses in a telescope through which we look out over the world. As those systems interact, they yield insights and nuances of understanding that no modality, acting alone, could produce.

Wide Angle and Narrow Angle Lenses

Optics systems enlarge their synergism by giving some lenses a broad field of view, others a narrow one. Along these same lines, in Figure 12-1 each successive conceptual system uses a wider-angle lens to look at life. Because the later systems are trying to cope with expanded complexity, they need a broad field of view to keep an eye on all the variables they face.

When we add wide-angle systems to the mix, narrow-angle systems, like System 2, gain an enriched perspective of existence. These less complex systems may retain simple models, but they no longer think of the world so simplistically. Thus, when paired with System 7, System 2 does not sport the same look that it does when teamed with System 3. Nor does System 3 have the same demeanor alongside System 4 that it does in partnership with System 5.

Scenes along the Mall

What we see, then, is that modalities change the way they evidence themselves, dependent on the other systems they interact with. To look at this in another setting, we might revisit the mall in Washington, D.C. You will recall that we were here once before, during our description of System 2 in chapter five. On our previous visit we made the point that this site serves as an equivalent of "holy ground" for our nation. It is where we enshrine the revered memory of illustrious and courageous compatriots.

One of the most frequented structures on the mall is the

Lincoln Memorial. Interestingly enough, the Lincoln Monument came into existence almost simultaneously with the closing of the American West and the acquisition of Puerto Rico and the Philippines, final chapters in creating the American colonial empire. With Indian Wars fresh in mind and jingoism still in the air, Systems 3 and 4 were dominant aspects of public life. Not surprisingly, the monument shows the influence of both those systems. It bears the massive scale and oversized statuary that typify System 3 empires and their capitals. To display their splendor and glory, empires have historically erected buildings of mammoth proportions and portrayed their leaders in larger-than-life statues.

The Lincoln shrine also carries the lines of classical architecture, itself a System 3 development. Viewed from afar the structure might easily pass for an ancient temple built at the zenith of Roman power. But this building is more than a tribute to Greco-Roman styles. Its bold symmetry speaks to System 4's esteem for order and stability. The words engraved on the inner walls are an eloquent statement of System 4 principles and values. Symbolically you must look up in order to read these noble truths etched in stone above your head. Thus, the Lincoln Memorial encapsulates System 2 hallowedness in System 3 and System 4 motifs.

Now position yourself a few hundred feet to the east at the Vietnam Memorial. Here we find no evidence of System 3 or 4 influence. Instead, System 2 is collaborating with Systems 6 and 7, modalities that were rapidly establishing themselves when this monument was conceived. Whereas the Lincoln Memorial seeks to create awe by overwhelming, the Vietnam Memorial achieves awe through intimate encounter. The statues do not stare down at you, as with Mr. Lincoln, but meet you face to face at eye level. They engage you personally. In keeping with the promotion of equality in System 6, statues commemorating female veterans were added later to the memorial. The famed wall, highly polished, compels you to see your own reflection as you gaze upon the names of fallen comrades. You become one with their tragedy. You even must kneel to read the names toward the bottom of the panels.

In keeping with System 7 preferences, the wall contours naturally into the earth, so much so that it escapes notice just a short distance away. And the statues are so unobtrusive that you hardly notice them at first. In your peripheral vision, you might easily mistake them for people in the crowd. Yet, for all its low-key understatement the Vietnam Memorial has become known as Washington's most gripping monument. The painful memories that still surround Vietnam partially account for this impact. But equally important may be the effective way the memorial unites System 2 reverence with newer systems that are starting to capture American minds.

System 2 is thus a significant contributor to the experience of both memorials. But the way it adorns itself in one instance is totally unlike the way it packages itself just a few feet away. This is the result of synergism.

A Player Goes Down

To illustrate our point in still another context, we might go back to the football game early in the chapter. As we sit in the bleachers among the fans, System 3 is prominent. Other systems are also present, often in a supporting role. One of these is System 2. With two tribes in relentless combat on the field, System 2 uses chants and cheers to unite each side in its fervor for war. Every now and then it fires up a fight song. Throughout the stands people clap their hands and stomp their feet to punctuate the rhythm, the equivalent of a tribal dance.

Now imagine that a player is suddenly and severely injured. Say a life-threatening neck injury occurs. Regardless of the team he is on, every voice goes silent. System 6 compassion immediately overwhelms the revelry of System 3. A pall settles over the entire stadium. All at once winning and losing are not so important. As the fallen player is loaded onto a stretcher, fans from both sides lay aside their rivalry to encourage him with united applause and cheers.

In that moment something intriguing transpires with System 2. No longer does it manifest itself the same way it did a few minutes before, in its supporting role for System 3. With System 6 at the helm, System 2 has an altogether different tone. It still claps and cheers, but with a solemnity that *unites* both tribes, not one that *polarizes* them. The applause amounts to a prayer, a petition to unseen powers to spare this young man's life.

Multiple Modes of Existence

Although our examples have emphasized how System 2 trans-
forms itself as it interacts with other modalities, all systems make
similar changes when they share the stage with another modality.
And usually multiple modalities are active at once. Indeed, one
advantage of referring to these systems as "modalities" is that we
can easily envision ourselves in multiple modes of existence simultaneously.

Think of what happens with Dad on a family vacation, for
instance. He probably spends hours behind the wheel in his
"driver mode." As he and Mom converse en route, he is also in his
"husband mode." All the while, he stays attuned to the children at
play in the back seat, being sure nothing gets out of hand. That
places him in his "parent mode." Each of these modes has differing priorities, and all of them are active simultaneously. What a
difference it would make if he were restricted to one mode at a
time, so that to talk with his wife he had to stop the van!

With the conceptual systems we do something similar. We
blend systems together in an appropriate response to the circumstances at hand. By doing so we expand our flexibility — the
very thing we require if we are to contend with a complex pattern
of ever-changing realities. We could not tap such flexibility if the
systems were absolutely autonomous and never joined forces.

By joining forces they also create a wonderful synergistic effect.
The result of this synergism is that any system can cloak itself in a
thousand disguises. Sometimes we have to observe a scene carefully before we can isolate all the systems that are part of it.

One reason we spent so much time detailing the characteristics
of each system in chapters five through ten is to help you become
that type of observer. As you begin to harmonize modalities in
your congregation, you will need subtle systems skills. A spiritual
experience that unites System 4 and System 6 will not look like one
that unites System 5 and System 6. And one that brings all three
together will look different still. Systems-sensitive leadership is at
the opposite end of the spectrum from one-size-fits-all planning. A
systems-sensitive church creates endless synergistic variety.

13

The Multi-System Church

Like the people who comprise them, churches have dominant modalities. This dominance may be strong or weak. It is strongest in those congregations that draw their membership almost exclusively from a single thinking system. The church then takes on that system's character across the board. Often this kind of church enjoys a natural consensus about how things should be done. Because people view things alike, they do things alike. Their collective action then shapes the church's ministry, organization, and style.

As systems diversify, this consensus breaks down. Where one modality previously held sway, others now rise to challenge it. Gradually the single-system church gives way to a multi-system church, with two or three modalities competing to set the agenda. In most multi-system churches there remains one dominant system, although it is no longer so strong as it once was. We do find cases, however, in which no one system represents a majority viewpoint, not even the one that was previously dominant. Instead, the church has strong, vocal minorities, each grouped around a particular system outlook and pressing to be heard. Consensus becomes problematic, particularly in the realm of methodology, for thinking patterns are worlds apart. Among other things, systems diversity will mean that members differ in

- how they learn
- how they organize committees
- how they want to be supervised
- what makes worship meaningful for them
- what helps them grasp a sermon

- what they expect of leaders, and
- what fulfills them by way of ministry to others

Not only do these differences exist, they are so pronounced that it is difficult to reconcile them. That is why multi-system churches experience strong centrifugal forces. It is also why tensions are growing more intense in contemporary congregations.

The Options Before Us

When tension starts rising because so many systems are present, congregations have only three alternatives. The first is to lapse into a laissez-faire existence. Unable to harmonize the competing viewpoints, leaders allow things to drift, hoping that somehow the underlying uneasiness will work its way out. This is a short-sighted approach, however. Laissez-faire leadership (if you can really call it leadership) ultimately yields aimlessness and demoralization. In these circumstances the church will either die of lethargy or ultimately end up in conflict. The conflict may stem from tension getting out of hand. Or it may break out when one of the modalities, fearful that the church is about to go under, asserts itself in an effort to rescue what remains.

The second alternative is to let competing systems slug it out. The unspoken object of this slugfest is for one modality to shed its minority status and establish its clear dominance. For that to occur, competing systems must disappear from the scene, either because they grow discouraged and leave, or because the "triumphant" system is able to run them off. To say the least, this hardly sounds like a solution Christ would endorse.

The third option is to hold the church together, yet keep it moving forward through systems-sensitive leadership. Unfortunately, this is not an enterprise in which leaders are generally well-versed, for the challenge of a multi-system church is relatively new. Until the last half of the twentieth century, System 4 was unchallenged as the dominant modality in most congregations. It determined leadership styles, sermon styles, architectural styles, educational styles — everything. And then suddenly, as if out of the blue, System 4 lost that absolute dominance. In little more than a generation Systems 5 and 6 became strong enough to rival it, with System 7 gaining rapid prominence.

We recently surveyed about 70 key leaders in a congregation to identify their dominant system. The results were

- System 3 — 1%
- System 4 — 45%
- System 5 — 22%
- System 6 — 22%
- System 7 — 10%

Although the percentages may vary, experience indicates that most urban churches, and many in smaller towns, share a similar distribution. Even if a congregation does not contain all these systems, odds are that the surrounding community does. When visitors enter our service, they are looking for evidence that we are sensitive to their dominant modality. To reach them and their friends, we must learn how to speak to and work with all the systems.

That is where systems-sensitive leadership comes in. It strives to nurture an environment that affirms, treasures, and encourages each system and its creative expression. It seeks to harness the inevitable systems tension so that it is productive, not destructive. The next few chapters offer "nuts-and-bolts" guidance on how to go about that. They examine almost every aspect of congregational life, showing you how to use systems insights to promote and maintain harmony.

Basic Strategies

Before we turn to specifics, however, we need to expand on the concept of systems-sensitive leadership and what we mean by that term. We have used it since chapter one, but have never elaborated on what it entails. This was not an oversight or some ploy to confuse you. Our purpose was to build a fundamental knowledge of the systems first, then look at specific strategies for managing them. Now that our systems survey is complete, it is important to map a fuller picture of what we have in mind when we talk of systems-sensitive leadership.

First, to state the obvious, it is a style of leadership. And leadership is in the business of anticipating what lies ahead. As leaders in general, and as systems-sensitive leaders in particular, we are trying to anticipate circumstances that will undercut morale, unity, and harmony. When we bring systems-sensitivity to our leadership, we add a broad dimension of insight that enlarges our ability to foresee and forestall conflict. We also know how to send multi-faceted messages to every thinking system, telling each of them repeatedly, "You really are treasured here."

Sometimes these messages are overt. More often they are subtle and indirect. We make no fanfare about them. We simply choose to do things with a certain style, to say things in a certain way, to structure congregational life in a certain manner, so that each modality feels affirmed and accepted. This is not manipulation, let us add. Instead, it is a decision to be intentional about something we cannot avoid. Everything a church does sends a signal, purposely chosen or otherwise, to one or more conceptual systems. They watch and react. Signals sent thoughtlessly are likely to be inconsistent and therefore confusing. Even worse, they may needlessly alienate and polarize. Systems-sensitive leaders try to avoid that pitfall by carefully managing both verbal and non-verbal messages. The aim is to assure goodwill in every modality.

A crucial vehicle for sending these messages is program planning. When systems-sensitive leaders lay out the work of a church, they create numerous islands of affirmation for every modality.

- They organize Bible classes around systems groupings, not age groupings.
- They assign teachers on the basis of a match between their natural style and the systems needs of their classes.
- They design worship experiences to accommodate the unique needs of each modality in the congregation.
- They prepare sermons with an eye to metaphors and themes that identify with each system.
- They are systems-sensitive in the choice of words and symbols that go into promotional messages and internal communication.
- They are attentive to architecture, decor, and landscaping,

aware of the non-verbal messages they send to various systems.

- They structure opportunities for hands-on ministry so that some are sensitive to the organizational preferences of one system, some to the preferences of another.
- Whenever possible they appoint ministry leaders whose dominant modalities align with the system characteristics of the ministries they lead.
- They build task forces and committees around people whose dominant systems allow them to work naturally together.
- Throughout the congregation they utilize extensive "feedback systems" to monitor reactions and feelings across the entire systems spectrum.

The purpose in all of this is to give each conceptual system a sense of congregational ownership. When System 4 walks in on Sunday morning, our goal is for it to feel, "This is a good place for System 4." But System 5 should walk in right behind it and say, "This is a good place for System 5." Even if a system is not congregationally prominent, we still should affirm its importance. For one thing, visitors from the community need to feel at home with us, and their dominant systems may not match those found among our members. If we do not connect with their dominant modality, they are unlikely to return. Additionally, in a group the size of a church new systems emerge quietly all the time. They, too, deserve a fellowship eager to receive them.

Pre-Empting Conflict

When we give each system a sense of belonging, then allay its anxieties and fears, we reduce the threat of conflict. Of greater importance, astute systems management builds genuinely healthy churches. It is this health, indeed, that diminishes the potential for discord. Or to put it another way, conflict avoidance is a primary benefit, not the principal objective of systems-sensitive leadership. We pursue systems approaches, first and foremost, to foster church health. In the process we create an environment that minimizes the risk of destructive tension.

But a caveat is in order. Systems-sensitive leadership holds out the prospect of *minimizing* conflict. It cannot promise to eliminate it. No matter how well we manage congregational systems, there will always be enough variables to give conflict an opening. Cynics would classify all congregations under one of three headings: those just *emerging* from conflict; those currently *embroiled* in conflict; and those about to *enter* conflict. While that assessment may be needlessly pessimistic, it does remind us that strife usually catches us unaware. We are not prepared for it, because we fail to see it coming.

We return, therefore, to the word "anticipation." Leaders must anticipate where trouble may arise, even when things are relatively calm on the surface. Even in strife-free churches, leaders should never presume smooth sailing ahead. They must always be watching and listening for the first tell-tale signs of danger. We are at a unique historical moment that demands such vigilance. The church is feeling its way through an unprecedented systems transition. As complexity increases, new conceptual systems are firing right and left. Change is sweeping us along like rafters in the midst of rapids, caught in a relentless current. We cannot paddle furiously enough to reverse course. We may long for the smoother waters upstream, but they are behind us forever.

The church, caught in the swirl of this change, faces countless systems challenges. Not the least of these is the management of change itself. We find ourselves at a juncture when new ways of doing things run into resistance, resistance turns to confrontation, and confrontation polarizes the church. In such an atmosphere it becomes incumbent on leaders to understand the dynamic of inter-system tension and know how to direct that tension into productive, not counterproductive channels.

A Smelting Pot Church

We might describe what we are trying to create as a smelting pot church. That is a rather unusual metaphor, we know. But once you understand its background, you will see its merit. The term came to mind while reflecting on a suggestion that William

Herberg made three decades ago. A sociologist who specialized in cultural diversity, Herberg questioned the practice of calling America a melting pot. His reservation grew out of observing vast dissimilarities in our land, the kind of thing we see in the great cultural divide separating a Cajun cop in Louisiana from a Scandinavian farmer in Minnesota.

That is not what we should expect from a melting pot, Herberg said. In a melting pot everything is liquefied and blended together. It all becomes one substance. What we resemble, he argued, is a "smelting pot." A smelter yields variety. From the same raw material it generates a host of products, each distinct in nature. An oil refinery is a good example of the smelting process. As it cracks a barrel of crude, the refinery bleeds off gasoline at one level, kerosene at another, naphtha at still another. Where a melting pot results in homogeneity, a smelting pot gives us *diversity*.

When Herberg advanced the "smelting pot" analogy, he had in mind social, ethnic, and regional diversity. But his analogy holds equally well with thinking systems. We are not all fashioned from the same mold. As a cross-section of the community around us, today's church looks more like the by-product of a smelter than a melting pot.

It is possible, of course, to ignore the smelting pot reality. Many leaders have chosen to do so. They simply continue to act as though their congregation is a melting pot, despite mounting evidence to the contrary. They insist on doing things in very specific ways, allowing for little latitude. But as time goes by, this type of authoritarianism proves increasingly unworkable. It only exacerbates the tension that is already present due to systems diversity. Then, as tension mounts, an autocratic mindset often sets in. "Our way or the highway" becomes the watch word. And frequently many do hit the highway, but only after an ugly rupture in the fellowship.

Diversity and Church Growth

Another response to the smelting pot reality has been the church growth movement. It has popularized the so-called "homogeneous unit principle." Fast growing churches, it holds, draw from a homogeneous element of society. According to this view, growth strategies should focus on a narrow niche of the community, then program purposefully, almost exclusively to that segment. This amounts to an intentional decision to put sharp limits on diversity.

In determining the segment to target, church growth specialists use ethnic and socio-economic categories. But they may have chosen the wrong vernacular. With your knowledge of the eight thinking systems you can see church growth strategies in a new light. The church growth movement is not targeting socio-economic groups. It basically targets conceptual systems.

There is just enough overlap between dominant modalities and some socio-economic units to permit this confusion. For instance, a church growth strategy aimed at yuppies will be laden with System 5 elements. That is because the yuppies think so dominantly in System 5 terms. Strategies aimed at older adults would include distinct System 4 programming, since System 4 is so prominent in that age group.

Questions about Homogeneity

While there are stellar examples of the homogeneity principle working (several mega-churches have grown out of it), many observers are uncomfortable with its exclusivism. Just because it works does not mean it is biblical, they argue. But while they reject the homogeneous unit principle, they frequently offer no equally promising alternative. They lack another model for building powerful churches in today's diverse and segmented society. We believe that a systems-sensitive church can provide that model. It is not our intent to debate the merits of the homogeneity principle, nor to judge its compatibility with Scripture. But we do hope to offer an understanding of diversity that paves the way for dynamic, multi-systems churches to emerge.

To borrow once more from our refinery example, churches might compare themselves to companies that sell petroleum products. Some specialize in only one by-product of the refining process. Perhaps they merchandise only to customers who are interested in propane or heating oil. Another might restrict its efforts to people wanting gasoline and lubricants, nothing else. Church growth literature advocates this type of congregation, one that segments the market and targets the customers within it.

We suggest a different strategy. It is equally fitting, we believe, for a church to see itself like a petroleum jobber who services customers whose interests cover the spectrum of petroleum products. The jobber is flexible enough to deliver what the occasion demands. One day it may be heating oil, the next day aviation fuel. But he is ready, as the apostle Paul said, to become all things to all people in order to win those he can.

A congregation that thinks like that jobber would be a "smelting pot" church. Attempts to build such churches in the past have met with only limited success, not because the basic concept is invalid, but because we did not bring systems-sensitive leadership to the enterprise. The "smelting pot" model offers an alternative to established congregations who have often found it impossible or too divisive to put the homogeneous unit principle into practice. When several systems already exist in a church, how can we adjust our focus and target only one or two of them? Who loses his right to feel at home in his own church?

That issue has troubled many churches who want to grow. In some instances congregational leadership has chosen to adopt the homogeneity principle even though it meant alienating long-standing members. One young minister took the position, "I'm willing to lose 500 people in the interest of building a church of 3000." He soon learned that while *he* was willing to pay that price, the church was not. A few weeks later he was gone.

An Alternative to the Homogeneity Principle

Other leaders, wanting to grow, but not willing to split the church in the process, have turned their back on programming derived from the homogeneous unit principle. But they have not known where else to look for building vibrancy and vitality in an established, but somewhat lethargic church. We believe systems-sensitive programming offers the model they need.

For the most part the homogeneity principle works most naturally in a newly planted start-up congregation. It does not conform at all to the leadership realities in an existing church, perhaps several decades old, that today finds systems diversity throughout the congregation. For churches like that, systems-sensitive leadership is a more promising approach. Not only that, if our conclusion is correct and the church growth movement is indeed targeting thinking systems, an inherent pitfall awaits churches built on the homogeneous unit principle. Namely, they will not be homogeneous very long.

Suppose a congregation targets System 5 successfully. Since we do not remain in a dominant modality forever, many of those early System 5 converts and adherents will move to System 6 or System 7 within a decade. At that point the church will lose its homogeneity and stress will start to rise. In short, even congregations founded on the principle of homogeneity will ultimately have to face the transition to a multi-system church. It may be wiser, therefore, to develop a smelting pot church from the outset.

Systems-sensitive leadership will not work miracles overnight, nor will systems skills be easily mastered. But if your experience parallels ours, systems sensitivity will start paying dividends immediately after you begin to apply it. That, in turn, will motivate you to make further effort at becoming a true multi-system church.

14

Learning to Accommodate New Systems

Quite frankly, systems-sensitive leadership will not come naturally to us. History has conditioned us to build churches around System 4 styles and methods. So strong is our conditioning that many of us cannot envision "doing church" any other way. Because we may feel uncomfortable if we depart too far from System 4 concepts of worship, organization, and ministry, efforts to accommodate Systems 5, 6, and 7 are frequently superficial. Our "accommodation" may amount to nothing more than minor tinkering that leaves the underlying System 4 structure unaltered.

This approach rarely works. It almost inevitably results in unprofitable forms of tension. On one hand stylistic change, unless managed carefully, can threaten members who characteristically think in System 4 terms. They fear it signals the first step in throwing out all "the old ways," including cherished truth. On the other hand we may be so cautious about change that Systems 5 and 6 question our commitment to their expectations. As a result, all three systems become irritated. Many a church now finds itself frozen in this very state of "dis-ease," with mistrust and misgivings steadily on the rise.

System 4's Historic Legacy

Why is it so difficult to get beyond this impasse? The answer lies in our unique systems heritage. System 4 has been the dominant intellectual force in Christianity since the days of the apostles. System 4 convened the great councils, hammered out the historic creeds, and gave us theological classics like Augustine's *City of God* and Calvin's *Institutes of the Christian Religion*. Even though System 4 shared power with other modalities for centuries, it always insisted on acting as their instructor. We can see this plainly in the Middle Ages. The medieval church drew its liturgy from System 2, its hierarchical organization from System 3. Yet sacrament and structure alike were answerable to the theological critique of System 4.

The Reformation fostered a major expansion of System 4 influence, especially in Switzerland. There reformers like Zwingli and Calvin replaced the System 2 trappings of medieval worship with System 4 exposition of Scripture. That pattern repeated itself wherever Calvinism spread. At the same time, System 4 also toppled the System 3 pyramid on which clerics had based their power. Luther delivered the first blow by insisting on the priesthood of all believers. Then Calvin completed the demolition by expanding lay leadership and making ministers answerable to their own congregations.

Soon all of non-Catholic Europe was experimenting with System 4 models for congregational governance. In effect, System 4 was consolidating its power. Having long held theological sway, it now took charge of how churches organized their work, perceived their role in the world, and conducted their worship. That situation would remain unchanged (although not always unchallenged) for more than 300 years.

Cloaking System 4 in Divine Mandate

As a result, modern churches learned to do business almost exclusively within a System 4 frame of reference. Public worship offers a classic example of that fact. Because System 4 felt a strong need for structure and control, spontaneity in worship became suspect in most mainline denominations and in many evangelical fellowships. Even churches that prided themselves on being "non-liturgical" conducted worship exactly the same way from week to week. System 4 also gave us the notion that the sermon should dominate the worship period. This reflects the priority that

System 4 places on inculcating eternal principles, working out their implications, and building life around them.

Had System 4 merely bequeathed its methods to the modern church, we would experience much less systems tension today. But history managed to cloak System 4 methods with the psychological equivalent of "divine mandate." That is, many people came to equate the methodology of System 4 with the way God *requires* us to do His work.

Back to the Bible

Two developments coalesced to produce this effect. First was the effort of the Reformation to elevate biblical authority. *Sola Scriptura* — "by Scripture alone" — was a watchword from the very outset of Luther's reforms. Heirs of the Reformation thus looked to God's Word to justify their doctrine and practices. Since the days of Luther and Calvin, most efforts to reclaim New Testament Christianity have pictured themselves as "back-to-the-Bible" movements.

Second, for generations this elevation of biblical authority was entirely a System 4 enterprise. Almost 400 years elapsed between the launch of the Reformation and widespread emergence of Systems 5, 6, and 7 in the general populace. During those intervening centuries, churches looked at Scripture almost exclusively through a System 4 lens. But this limited systems view had a telling effect. Anytime we study the Bible through the eyes of a single conceptual system, we become overly sensitized to that system's themes. They stand out boldly as we read. Conversely, we skip lightly over themes related to other modalities. We may even miss those concepts altogether.

And that is precisely what happened in the wake of the Reformation. Passages that endorsed System 4 styles of worship and ministry gained ready attention. Time and again they were the anchoring text for sermons and doctrinal studies. By comparison, non-System 4 approaches to spirituality — especially those related to Systems 5, 6, and 7 — lay unappreciated on the page.

For many people, as a consequence, "doing Bible things in Bible ways" became synonymous with doing them in System 4 ways. System 4 methods were not merely *acceptable* to God. According to this mindset they were the *only* acceptable way to serve Him. From hindsight it is easy to fault such exclusivism, but we should do so judiciously. Until Systems 5, 6, and 7 became ascendant, System 4 was the only intellectual modality around. It alone promoted literacy, built educational institutions, and collected libraries. It therefore dominated the conceptual landscape. To do theology at all, people had no choice but to use System 4 thinking patterns. Restricted to a single mode of conceptualizing, it was easy for Christians to assume that the models they derived from Scripture were the only legitimate ones.

Rivals to System 4

Not until we approached the twentieth century did other intellectual systems rise to wholesale levels among everyday people on the street. In time these emerging systems would challenge the System 4 consensus in every discipline, including religion. Churches, however, were among the last institutions to feel pressure from these newer thinking patterns. The earliest waves of System 5 entrepreneurs and System 6 care-givers were willing to work with (or even preferred) a System 4 church. In their professional life they might function in Systems 5 and 6, but when they entered a worship service, they shifted to System 4. As a result, churches became complacent. They presumed themselves immune to the change virus that afflicted others.

The 1970s and 1980s completely dislodged that complacency. As television came into its own, it besieged every American home with a barrage of System 5 and System 6 themes. Over those decades viewers increasingly appropriated System 5 and System 6 values for themselves. At the same time, the American labor force shifted from blue collar workers to professional and technical specialists. With that transition the workplace, historically governed by System 3 and System 4, transformed itself into a setting where Systems 5 and 6 prevailed in the most lucrative and influential career fields. Consequently, churches woke up one day to find bright, capable, highly educated System 5 and System 6 thinkers in every pew. In some places they constituted the congregational majority.

In addition, these newer systems were no longer taking a passive role. They had emerged as a voice for thoroughgoing change. Today, as their numbers grow inexorably larger, that voice is becoming stronger. They call for a different style of worship, a different way to organize the church's work, and a larger role for women, just to name some of the more volatile issues. All of this is quite disturbing for System 4, which, while not opposed to change, wants to proceed cautiously with it. System 4 also wants definite boundaries on change. It is more likely than Systems 5 and 6 to draw lines and say, "No matter what, we cannot change beyond this point."

Diverse Views of Worship

To illustrate how tension develops between System 4 and these other modalities, we might look at the subject of worship. We have already mentioned that System 4 prefers a Sunday morning service that is predictable and controlled. System 4 also thinks of the worship hour as basically a vertical experience in which believers turn their minds heavenward through praise and prayer. One System 4 church has a banner atop its program of worship that reads, "Enter quietly. Reflect silently. Worship reverently." Only this type of hushed, reflective atmosphere connotes what System 4 defines as reverential respect.

Given their way, Systems 5 and 6 would have a worship service that is somewhat noisy and certainly informal. These systems emphasize the horizontal aspects of the Sunday experience. As with System 4 they value worship that renews dialogue with heaven. But they also see worship as a time to strengthen the bonds of community among believers.

The Loss of Community

This emphasis grows out of the unique conditions that foster the appearance of Systems 5 and 6 in a culture. These systems gain prominence only in highly complex societies. One price for that complexity is the loss of relationships. System 2 thrives on a sense of tribe, System 4 on a sense of neighborliness. Systems 5 and 6, however, lack a natural group with which to relate. By the time System 5 becomes socially dominant, tribes have disappeared, replaced by nuclear families, and neighbors no longer know each other's names. Thus, Systems 5 and 6 experience an unprecedented level of isolation, even in the midst of a sprawling city.

Their isolation carries over to their church life. Prior to System 5, worship was merely another gathering of family, friends, and neighbors. People who worked and lived together came together once more on Sunday for worship. The church did not have to worry about building member-to-member relationships, for they already existed. Not so in a System 5 society. System 5 and System 6 worshipers are as unlikely to know their neighbor on the pew as their neighbor down the street.

To regain a sense of connection with others, Systems 5 and 6 create "forged communities." By "forged community" we mean one that does not occur naturally. Neither shared bloodlines nor a shared neighborhood produce it. Instead, it holds together solely on the basis of common interests. Remove those interests and the relationship goes away.

In this scheme of things, the church becomes a forged community for Systems 5 and 6. They often drive miles to the church they attend. Their only relationship with others in the congregation may be the church itself. More than likely none of their neighbors or work associates attend. For that matter, neither do family members outside the immediate household. Except for times of worship, Systems 5 and 6 may never see another member of their church.

A Church That "Feels" Like Family

Even when they do find friendship in a congregation, those friendships may not form immediately. And sometimes they never form at all. In the meantime, System 5 and System 6 can attend worship week after week and feel absolutely cut off from the people around them. They need the church to help them build a sense of community with other believers. They want church to "feel" like family. Indeed, one sign that System 5 and System 6 are emerging in a congregation is the marked elevation of "family" as a metaphor for the church.

To Systems 5 and 6, however, a worship service does not "feel" like family if it is formal, controlled, and subdued. These systems view traditional worship assemblies as too stiff, too solemn, too lacking in energy. "Families" simply do not operate that way. System 5 therefore pushes for great variety and spontaneity in worship. If the format varies from week to week, System 5 finds the change of pace refreshing. In addition, System 5 wants people to "connect" with the service. It shuns anything that would impede understanding or participation. To that end it promotes contemporary music, readings from conversational translations, and prayers that avoid Elizabethan language. It also wants sermons that are practical, down to earth, and packed with "how-to's" for daily living.

Informality

System 6 calls for even greater informality. It believes that the church should gather on Sundays as much for fellowship and mutual encouragement as for worship. System 6 points to the "one anotherness" texts in Paul's letters as a guide for what should happen when Christians meet. It wants a worship service that throbs with interpersonal energy. Or to borrow another image, System 6 loves services that evoke the feeling of friends enjoying themselves around a dinner table. Preaching, in the estimation of System 6, should emulate conversation in a family circle, replete with story-telling, candor, and lessons shared from personal experience.

Now, neither System 5 nor 6 wants to strike the vertical component from worship. That is not their desire at all. Some of their favorite songs are hymns of praise and adoration. But what they seek is praise anchored in a family-like setting — warm, intimate, and interactive, with extensive lateral communication, perhaps at the expense of sermon time.

For instance, many congregations now pause during the Sunday service to let worshippers introduce themselves to others nearby. This practice grew out of System 5 and 6 promptings.

Dress codes also relax as Systems 5 and 6 become prominent, especially System 6. This reflects the emphasis on informality that is at the heart of System 6 values. As a rule of thumb, the more System 6 you find in a congregation, the fewer coats and ties you will see on Sunday morning. (There are, of course, exceptions in places like California and Hawaii, where dress codes are generally relaxed across the board.)

While such changes seem minor to Systems 5 and 6, they may disturb System 4 deeply. Once worship begins, System 4 does not want the noise and hubbub of people interacting with one another. From a System 4 perspective "greet-your-neighbor" activities are a distraction, a diversion from the God-centered purpose of the hour. Moreover, System 4 may want limited interaction not only *during* the service, but even *prior* to it. In the words of that program of worship we mentioned above, System 4 wants people to enter quietly, sit reflectively, and prepare their minds for what follows. To System 4, anything less reflects a superficial attitude toward worship. System 4 may also censure relaxed dress codes in worship if it views casual attire as a lack of respect for God.

Unheralded Exits

With such differing perspectives, it is easy to see why tension develops over styles of worship. We could easily illustrate this same systems tug-of-war in a dozen other areas of church life. Tension often comes to a head when leaders accede to entreaties from Systems 5 and 6 for change. If those changes tread on System 4 sensitivities, System 4 lets everyone know about it. System 4 buttonholes leaders, circulates petitions, and writes anonymous letters more quickly than other modalities. When System 5 or System 6 are offended, they are likely just to leave. We look around one day and miss them. We suddenly discover that they have either quit coming or have placed membership elsewhere.

They can make this unheralded exit because it is easier for Systems 5 and 6 to cut their ties and leave a church than for System 4 to take that course. Going back to chapter seven, you will remember that System 4 forms its sense of personal identity around the organizations it is part of. Wherever it manifests itself, in both secular settings and religious ones, System 4 tends to denominationalize around the prevailing ideology. It then build its sense of "who I am" on that denominational affiliation. I am a Democrat; I am a Republican. I am a Methodist; I am an Anglican. I am Sunni; I am Shiite. Or in Marxist circles, I am a Leninist; I am a Maoist. Just as dyed-in-the-wool Democrats feel they have betrayed a fundamental loyalty if they vote Republican, System 4 Christians are immediately guilt-struck if they walk away from their congregation. That church is essential to their identity, to their sense of who they are.

Non-Judgmental Exits

By way of contrast, Systems 5 and 6 look at organizations altogether differently from System 4. Systems 5 views the church, as it does all other institutions, as a place where it should be mentored and made more effective. Except for those enterprises or companies it personally creates, System 5 rarely binds its identity to an organization. Thus, if System 5 is in a church where its spiritual needs are not addressed, it simply moves on. It feels no guilt nor hesitation in doing so. Nor does it leave with rancor. It just leaves.

All of this came home tellingly one day during lunch with a talented young professional, himself the epitome of System 5 values. He and his family had left our congregation some months before. Immediately after we met in the restaurant, he started asking about dozens of people at church, how they were doing, what they were up to. He was obviously very interested in these people and had a deep affection for them.

But it was equally obvious that he would not be back in our services. He had taken his spiritual quest elsewhere. When he talked about his decision to leave, he said in effect, "If you want to do things that way at your church, you certainly have the right to do so. I don't have any problem with that. It just doesn't connect with me. So I'm going to find a place that speaks to my own spiritual concerns."

The same thing happens with System 6. If System 5 is prone to pack up its tents and slip quietly away, System 6 is doubly prepared to do so. System 6 is suspicious of organizations to begin with. It believes that organizations have no soul, that they lose their heart and compassion in trying to perpetuate themselves. Anywhere you see System 6, it is probably working to minimize institutional structure. Its sense of "who I am" thus has little if any connection to a particular congregation of Christians. It may like them, even love them. But it will readily leave them.

Systems Perspectives on the Church

Does the ease with which they leave mean that the church does not matter to Systems 5 and 6? Not at all. They simply do not evaluate a church the same way System 4 does. This is not because Systems 5 and 6 are indifferent to God's calling or to the importance of His church. They may in fact be quite serious about both. But because they have donned other systems lenses, they do not look at the church the same way System 4 does. They have a different perspective on what the church should be about.

To understand this shift in perspective, we must recognize how we as Christians form our personal opinion of a congregation. Deep in the core of our being we experience a profound sense of divine expectation. It may come from our study of the Bible, from the training of our parents, or from years of spiritual reflection. In any event, we turn to the church to help us respond to this expectation. We judge a congregation on the basis of how well it helps us conform to our inner sense of what God calls us to do.

But this inner imperative, at the core of our existence, contours itself differently as we move from one modality to the next. It is always a divine mandate. But in System 4 it is the mandate to live morally; in System 5 to live effectively; and in System 6 to live selflessly. Therefore, System 4 judges a church by the standards it upholds, System 5 by the spiritual skills it imparts, and System 6 by the service it gives to hungry and hurting people.

System 4, we must hasten to add, also cares about effectiveness and compassion. But it cannot imagine how deeply these concerns shape the very soul of Systems 5 and 6. When System 5 looks to the church, it asks for sustained guidance on how to be a better parent, a better business person, a better leader in the community. System 4 may sees those as "nice-to-haves," but hardly essentials in the church's teaching program. System 6 asks for regular opportunities to be in the frontline battle against poverty and human misery. System 4 may answer, "Yes, we give money to those types of causes," but sense no urgency to create hands-on experience for System 6 to relieve human want.

Yet, to hold onto Systems 5 and 6, a church must take those requests seriously. Otherwise it leaves these systems spiritually frustrated and unfulfilled. They will feel they are failing God in the very thing He has called them to do. They can no more be at ease with that than System 4 can make peace with a morally indifferent church. System 5 and System 6 may see the congregation as a good one, filled with good people. But they will no longer see it as the place they need to be.

Doctrinal Particulars

In finding a place that addresses their spiritual needs, Systems 5 and 6 may even opt for congregations with which they disagree in several doctrinal particulars, something System 4 almost refuses to do. System 5 sees its salvation as dependent on its own individual faith and response, not on the purity of the congregation it happens to associate with. System 5 considers itself capable of discerning truth and holding to it, even if surrounded by others with totally different views. It can attend a premillennial church without thinking of itself as premillennial. It can be part of a charismatic church without thinking of itself as charismatic. Its identity is not tied to its church affiliation. As a consequence, doctrinal alignment is not an absolute prerequisite when System 5 (or System 6, for that matter) chooses a congregation.

Both System 5 and System 6 normally look for substantial doctrinal overlap in a new church, to be sure. But once they find that overlap, they do not preoccupy themselves with the doctrinal differences that remain. Instead they begin asking how this congregation can advance their own spiritual journey, how it can foster their Christian development. When they answer those questions to their satisfaction, Systems 5 and 6 are ready to drive down new tent pegs. As a footnote, we should add that System 4 cannot understand this willingness to cross doctrinal lines so freely. When it sees this happen, System 4 wonders if System 5 and System 6 ever loved truth in the first place.

Even though we have stressed the relative ease with which Systems 5 and 6 change churches, we are not suggesting that System 4 will stick around forever. It, too, will bolt. But normally it takes its departure only after prolonged and arduous struggle. System 4 is vocal in its discontent. It never steals away quietly, the way Systems 5 and 6 sometimes do. Because its own identity is so tied to the church itself, changes in congregational style can pose a threat to System 4's personal sense of identity. That is why System 4 starts buttonholing leaders the minute it becomes upset. It feels personally threatened. As soon as we offend System 4 sensitivities, we know about it. With Systems 5 and 6, on the other hand, we may never realize the degree of their discontent until they are gone.

15

The Primary Systems Conflict

As we made our way through chapter fourteen, perhaps you noticed that we narrowed our focus to Systems 4, 5, and 6. There was a purpose in doing so. While no two systems get along perfectly, these three create the greatest tension in the church today. On some occasions our problems come from differences between System 5 and System 6. But the most common struggle is between Systems 5 and 6 on one side, System 4 on the other.

By nature System 4 is cautious, conservative, and guarded. It perceives itself as the protector of truth and principle. It is therefore slow to abandon proven ways until it feels assured that new initiatives will not sacrifice vital norms. Systems 5 and 6, by contrast, are innovative and spontaneous. They push things farther and faster than System 4 thinks prudent or even necessary. When they do, System 4 feels duty-bound to put on the brakes. Its own sense of responsibility gives it no other choice.

Systems 5 and 6 then respond to System 4's reticence by charging that it is rigid, entrenched, and relentlessly opposed to change. While that indictment is sometimes accurate, often it is not. System 4 *will* support change, but only if it trusts the change agents. Ironically, both System 5 and System 6 tend to pursue change in ways that, far from securing System 4's trust, actually work to undercut it. (We will see examples of this later.) And what System 4 does not trust, it both fears and opposes.

Moreover, System 4 is close enough to System 3, and it has warred with System 3 long enough, that it fights with tenacity when provoked or frightened. The result is an impasse that many church leaders know only too well. On one side of the aisle

System 5 and System 6 are calling for new styles of worship and ministry. On the other side System 4 is digging in its heels, afraid things are getting out of control. As a rule both sides are doing their share of finger-pointing and name-calling. "You're closed minded and inflexible," Systems 5 and 6 yell at System 4. "You don't care about principles and basic values," System 4 shouts back. And leaders, caught in the middle, are worn out and frustrated by the entire ordeal.

The church thus faces a singular systems challenge. We must find a way to make changes that will accommodate the legitimate spiritual needs of Systems 5 and 6, yet do so without engendering System 4's distrust. At present this is the most difficult systems issue in the body of Christ. Fortunately it is possible for these three systems to work hand-in-hand, even if they cannot see eye-to-eye. Instead of a confrontational atmosphere, we can move to a win-win situation for everyone. Moreover, we can get there without abandoning Christian fundamentals or basic biblical values.

Flash Points

No fixed rule allows us to anticipate the rise of systems tension, nor to predict its intensity when it arises. There are notable variables, however, which serve as generally reliable predictors.

- Tension sparks conflict most readily in congregations that have one or more unhealthy systems. (We will talk about signs of an unhealthy thinking system at a later point.)
- Measured change creates less conflict than sudden change.
- Change made necessary because of a disaster (the church building burns down, the local economy collapses, etc.) is far more palatable than change arising from other factors.
- Changes to accommodate Systems 5 and 6 create less tension when those initiatives receive visible leadership from someone System 4 thinks of as "one of our own."
- The more rapidly Systems 5 and 6 become prominent, the more intense the tension.

Certain scenarios are particularly prone to systems clashes. The common denominator in many of these is a rapid influx of new members. If the newcomers come from one set of dominant modalities, the long-established membership from another, the odds for conflict multiply. Here are some typical instances in which this occurs.

- After relocating its facility, a church starts drawing from neighborhoods whose socio-economic makeup is altogether different from its historic membership base.

- A new factory or corporate headquarters moves into town and brings a flood of new employees to the community.

- The congregation hires a highly popular and personable preacher who begins to attract a host of new members.

- The church picks up dozens of families who are fleeing conflict in another church in town.

- Young families, working in a nearby city, move in droves to a rural community to find a lower cost of living.

- A small town becomes snared in the sprawling expanse of a fast-growing suburb.

When conflict arises in these types of settings, we typically assess it as an old-guard/new-guard rivalry. And the dividing lines often fall out that way. But closer examination reveals a systems conflict in which the old guard aligns by happenstance with one modality, newcomers with another.

Systems Tension: A Case Study

To demonstrate what we mean, consider a rural church like the one we mentioned above, caught in the path of encroaching suburbs. From a systems standpoint, this church is probably in transition from System 3 to System 4, most likely toward the end of that transition. In other words, its outlook is dominated by System 4 perspectives, but with strong supporting System 3 influence. This profile is common in rural churches and in small town America, where a blended culture of System 3 and System 4 still has deep roots.

People moving into the church from nearby suburbs probably come from a different systems mix. Suburbanites, especially younger ones, are typically given to high levels of System 5 or

System 6. Despite that fact, the church may ignore this systems difference at first. It welcomes these newcomers with open arms, because it recognizes a need for new, young blood. Trouble begins only later, after Systems 5 and 6 become a sizable presence and no longer consider themselves "new members." At that point they will press for a voice in decisions, as well as new directions for the congregation.

They may not get that voice, however, much less the new direction they want. Leadership in System 3/System 4 churches tends to be self-perpetuating. That is, those who currently lead are able to hand pick any others who join their ranks. In this type of setting Systems 5 and 6 may grow large, or even become a majority in the pew, yet remain unrepresented in decision-making circles. They are kept at bay by a leadership that may not be able to put its finger on the problem, but somehow feels uncomfortable with these newer elements.

Asking For a Voice

Believing they have right to be heard, Systems 5 and 6 will not accept this state of affairs forever. They will eventually expect a forum in which to voice their views. If nothing else, they will ask to be consulted when major decisions are made. System 3/System 4 leadership, on the other hand, does not have a consulting style. More often than not it takes counsel only with its inner circle, makes its decisions privately, and tends to act unilaterally.

Moreover, the congregation has always preferred this very type of leadership. System 3/System 4 churches like leaders who are decisive, who have everything under control. System 3/System 4 church members feel no need to be consulted. From their perspective the ones who lead know what is best. After all, that is why they were chosen to lead.

System 3 and System 4 find it unsettling, therefore, when Systems 5 and 6 mount a push to be heard. First of all, this pushiness smacks of insubordination, a serious offense in any System 3 or System 4 organization. In Systems 3 and 4, leaders decide and followers follow.

Second, these newcomers advocate a host of ideas that strike

System 4 as ill-advised, even dangerous. The present way of doing things has served the congregation well for decades. Why start tinkering with it? "If it ain't broke, don't fix it," System 4 says. And System 4 sees nothing in this church that is seriously broken. After all, it is growing rapidly, perhaps more quickly than anytime in its history. Leadership cites this as proof that their way of doing things is working, and working quite well. (What they overlook, of course, is that sheer demographics are driving their growth.)

A Tactical Mistake

While trying to settle on a response to Systems 5 and 6, the leaders get plenty of unsolicited advice from System 3/System 4 members who are worried about all the talk of change. This portion of the congregation has a singular message for leadership: "Stand firm and stand pat."

Because the leadership is new to this type of situation, it often slips into a tactical error. Not knowing what to do with the voices for change, it turns an increasingly deaf ear toward them. It presumes, wrongly, that System 5 and System 6 will be compliant followers, like those it has always dealt with in Systems 3 and 4. It therefore thinks a simple "no" to Systems 5 and 6 will suffice to calm things down.

To leadership's surprise, System 5 and System 6 respond by pressing their case more fervently. Now leadership not only does not listen to Systems 5 and 6, it begins to view them as rebels — as System 3 in clever disguise. It therefore redoubles its effort to hold them at bay. Above all it feels compelled to deny them an opportunity to gain control. If leadership was hesitant to give them a role in key decisions before, it is determined not to do so now. At this point, lines are becoming sharply drawn. Tension is clearly mounting. Unless something dramatic occurs, it is only matter of time before there is a wholesale walkout or a congregational blowup.

It's Not Generational

It may seem that we have cast System 4 as the "heavy" in this case study. But that is merely coincidental. We could choose other situations in which System 5 or System 6 would be treating System 4 heavy-handedly. We merely wanted to underscore that conflict like this has more subtle underlying themes than simple rivalry between the old guard and the new guard. What we are dealing with is the type of systems tension that is always waiting in the wings whenever Systems 4, 5, and 6 come together in a church.

In many instances we confuse System 4's conflict with Systems 5 and 6 as an intergenerational struggle, i.e., as the older generation versus the younger one. This, too, is an inadequate analysis. But because it is so common, we need to devote a few comments to it.

When writers and researchers accredit tension in the church to generational struggles, they usually make the dividing line in this standoff somewhere around age 50. They portray those in their fifties and above as having one set of outlooks, those in their early forties and below holding to a worldview that is altogether different.

That would place those born before the Second World War in one camp, those born after the war in another, with the people born during the war years serving as a transition group. This approach has given rise to the common practice of contrasting the mindset of baby-boomers and the so-called "baby-busters" to the values of their parents and grandparents.

While this simplified model of the conflict has its merits, its very simplicity distorts what is happening. We are dealing with a systems struggle, not generational one. It just so happens that there is a high concentration of System 4 dominance in the generation born before the Second World War. It also happens that a similarly high concentration of System 5 and 6 thinking appears in the boomer and buster generations. That is why it is so easy to conclude that we face a generational problem.

But if church leaders think of today's uneasiness in the church as merely generational strife, they miss the truth of what is happening. When we examine congregations closely, we find many older people, often in their seventies or eighties, who are striking examples of System 5, System 6, and System 7 thinking. Conversely, we find strong pockets of System 4 dominance among young adults.

In fact, over the past two or three years we have encountered several instances in which church leaders, making changes to accommodate Systems 5 and 6, ran into unexpected resistance and rigidity from younger adults. One congregation, considering a contemporary format for its worship, discovered that many in their twenties opposed the change. Their comments indicated they had chosen the congregation because its worship *was* traditional. They did not want a System 5 or System 6 church. They were looking for something that embodied System 4 values and outlooks.

Boomers, Busters, and NDOs

As yet there are no detailed studies of this resurgent System 4 dominance among younger church members. We therefore do not understand it fully. But permit us to offer some impressions based on encounters with this phenomenon for almost four years now. First, as we said above, the resistance to change among adults in their thirties and forties is itself surprising enough to leaders. But when that resistance is also quite rigid, leaders are doubly taken aback. We should hasten to add that this rigidity is by no means universal. Boomers and busters who are System 4 dominant are frequently quite flexible. Yet there are notable exceptions.

In thinking about the exceptions we have encountered, we have noted a prevailing pattern. Namely, these more inflexible System 4 boomers and busters tend to come from among a group we would label the NDOs, the "never-dropped-outs." In an unprecedented historical development, two-thirds of the baby-boom generation dropped out of active church involvement when they left home. By middle age about half of those returned. But just as many remained on the sidelines. The NDOs are that one-third of the boomers who never left. In cases where their children have also stayed active in the church during early adulthood, those children, too, are NDOs.

System 4 Boomers and Busters

Our theory is that the rigidity we sometimes find in NDOs is a reaction to what happened with their non-religious peers. Unlike those peers, who were decidedly System 5 and System 6 in outlook, the baby-boomers who never left probably had a strong System 4 makeup to begin with. Otherwise they would have joined the exodus when so many of their friends walked away from the church once its System 4 style no longer communicated with their needs. Those NDO boomers who have remained System 4 dominant (not all have, by any means) typically gave their children a strong System 4 upbringing. Those patterns of thought in their children now reflect themselves in the baby-busters who are System 4 dominant.

The NDOs, both boomer and buster, have witnessed firsthand the moral collapse of our nation. They have watched their generation addict itself to drugs, divorce in wholesale numbers, and become hedonistic to the core. For NDOs who are System 4 dominant, that has convinced them that "the old ways are the right ways." Even though there are incidents of drugs and divorce among NDOs, it is nothing like what is happening elsewhere in their generation. That is why some NDOs cling tenaciously to traditional religious expression.

Intriguingly, many of these very NDOs who want a System 4 church are also given to System 5 and System 6 characteristics in their social and professional life. They drive the same status automobiles, send their children to the same prestigious colleges, and fill their homes with the same techno-miracles as their System 5 neighbors. They may be involved with ecology efforts, recycling programs, and projects to care for the homeless, just like System 6. But when they come to church, they want an experience that is distinctively System 4 in nature.

A Resurgence of System 4

In working with this element of the church, one of our first objectives is to help them develop healthy System 4 thinking. Rigidity always indicates that the dominant modality is unhealthy, whether we are talking of System 4 or some other. Most System 4 thinking is reasonably healthy, whether in the young or the old. Where it is not, we must work to rectify that condition. We will have more to say on this subject in chapter twenty-four, where we look at the factors that give rise to unhealthy systems.

Our point here is that there is a resurgent System 4 dominance among many young adults that we must not overlook. It is one of the reasons that conservative talk show hosts have a vocal following among adults in their twenties and early thirties. System 4 in the baby-busters may be a natural reaction to the permissiveness and lack of strong boundaries they saw in their baby-boomer parents. Perhaps it is taking the baby-busters a longer time to work through System 4 issues because they did not have a strong System 4 structure at home or in school during their formative years. Whereas boomers went hurriedly into Systems 5 and 6 as a reaction to their System 4 upbringing, the busters had comparatively little System 4 guidance in their upbringing. They are on their own as they feel their way through the resolution of System 4 issues.

If we interpret tension in the church as merely intergenerational strife, we overlook the systems diversity that crosses all age lines. If we target the busters exclusively with System 5 and System 6 programming, we may miss their System 4 needs in wholesale lots. Similarly, to target only System 4 programming at older adults may prove just as short-sighted.

One beauty of systems-thinking is the promise it holds for intergenerational activities. When we plan congregational functions and processes to appeal to specific systems rather than specific generations, an amazing thing happens. Those who are System 6 dominant, both young and old, are drawn to the same activity. The same thing happens with the young and old from System 4 and System 5 when we build systems-specific events. The "birds-of-a-feather" principle emerges anew, this time in systems garb. And suddenly the generations are working side by side again in the church.

We Need Every System on Board

Despite the natural tension between Systems 4, 5, and 6, these systems *can* coexist peacefully and productively. If we were not convinced of that possibility, we would hardly have wasted time

writing this book. In the pages ahead we offer specific suggestions
by which systems-sensitive leadership can help these modalities
rise above in-fighting and rivalry. Achieving such harmony is critically important, for the work of the church needs the contribution
of every system. By now we have seen that each modality has its
strengths, each its weaknesses. When all the systems work hand-in-
glove, they complement each other. The strong points of one
offset limitations in another.

We need System 4's regular reminder that we cannot relegate
truth and basic morality to a subordinate position in the Christian
value system. We also need System 5 to keep prodding us to do
things better, more effectively, and to help each individual achieve
his or her God-given potential. Then, too, we need System 6 to
keep us focused on people who are hurting, on the priority of
compassion in the ministry of Jesus, on the need to be a healing
presence in relationships all around us. Nor should we overlook
System 7 and its contribution. We need its focus on the future, on
sweeping changes that are just over the horizon, along with its
creativity in helping us anticipate what we must do to be ready for
those changes.

Our goal must be to capitalize on the strengths of every
modality and create a win-win situation for everyone. Moreover,
we can get there without abandoning Christian fundamentals or
basic biblical values. Still, getting to win-win will not be easy. It will
require two vital developments. First, each system must set aside
its arrogance. There is just enough ego in the individual systems
for them to say, "We wouldn't have any trouble around here if
everyone was smart enough to do things my way."

Learning Forbearance

Second, all parties must want harmony, or none will prevail.
The various systems will always see things differently, enough so
that someone determined to set them at odds can usually achieve
that purpose. Therefore, the first duty of systems-sensitive leadership is to instill a love for unity throughout the congregation. We
will have more to say on this subject later. We mention it now only
to highlight an absolute essential. With systems tension inevitable

in the years ahead, we can only sustain unity in a setting where forbearance is practiced across the board.

Please note that we use the word "forbearance," not "tolerance." The choice is intentional. Although you will often hear tolerance described as a Christian virtue, it may not deserve the title. The one time the verb "tolerate" appears in the New Testament, the connotation is clearly negative. In His words to the Christians of Asia, Jesus accuses the church in Thyatira of "tolerating that woman Jezebel" (Revelation 2:20). There was not enough concern about her conduct, in other words, to take action against it.

Which brings us to the problem with tolerance. Being tolerant does not necessarily indicate a magnanimous spirit. Instead, it may be merely a sign that we are totally unconcerned. Indifference alone will permit us to tolerate something. We tolerate pervasive pornography because we are not upset enough to stamp it out. We tolerate crime in the streets because we are not perturbed enough to stop it. In a word, we can be passive, detached, and disinterested, yet still have a reputation for being tolerant.

Forbearance, on the other hand, is never passive. It requires us to "bear for" others what they are unable to carry by themselves. Forbearance requires us to become involved, to help shoulder the load. When Paul encourages the Ephesians to show forbearance to one another, he entreats them to attend that effort with humility, gentleness, patience, and love (Ephesians 4:2). Today, more than ever, we need to emblazon that reminder above every church. As we shall see in chapter sixteen, Systems 5 and 6 have an uncanny knack for pushing the fear buttons in System 4. We will also find that System 4, when it becomes fearful, is quick to impugn System 5 and System 6 motives, often without merit. For systems harmony to succeed, all parties must learn new ways of relating to each other.

16

Understanding Resistance to Change

When churches fought a hundred years ago, the issue was almost inevitably doctrinal. Not much had changed in that regard since New Testament times. For 1900 years doctrinal differences accounted for most fellowship rifts. But that is no longer the case. We still have our share of theological squabbles, to be sure. Yet they increasingly take a back seat to a new kind of polarization that has emerged in the last half-century, cutting across all denominational lines.

Today the most persistent and disabling congregational tension often centers on issues of style. Disagreements about the type of songs to use in worship. Dissatisfaction with the way leaders reach and implement decisions. Clashing expectations for the youth program. Discontent with the pace of change — in one quarter because change is too fast, in another because it is too slow. Some urge the church to become more contemporary, for it risks losing relevance to its culture. Others fear that the church has already changed so much that it is in danger of losing its moorings.

These kinds of disagreements are increasing because so many systems now share the pew. Stylistically and spiritually each system has its own preferences and priorities, and it clings to them dearly. Not only that, every modality can cite what it considers adequate biblical support for "doing church" its way. Arbitrating these disputes is therefore difficult.

Our purpose in this chapter is to identify tactical errors that Systems 4, 5, and 6 must avoid if they are to work together peaceably. These mistakes occur naturally. They result from instinctual differences within the modalities themselves. Space will not permit us to examine all these differences, but we will take time for the ones which contribute most commonly to inter-system strife.

Views of Human Nature

To begin with, System 4 is far more pessimistic about human nature than Systems 5 and 6. Yet that pessimism is hardly ill-founded. System 4 comes into existence as a reaction to System 3. Or to put it another way, System 4 defines itself as a counterbalance to the pleasure-seeking, sometimes ruthless lifestyle of System 3. Having devoted so much energy to restraining System 3 impulses, System 4 has been conditioned to view human nature with misgivings.

System 5 emerges, by contrast, only after System 4 has established a firm grip on System 3 and brought it under control. Once in command, System 4 promotes what is noble and altruistic in humanity. It brings justice and fair play to bear on every circumstance. It builds strong, orderly communities that promote virtue, duty, and loyalty. Thus, System 5 comes to the fore in a setting where men and women are at their best morally, not at their worst. System 5 is so far removed from the undisciplined world of System 3 that it does not identify with the threat that System 3 can present. Instead, System 5 thinks of human nature as given to the stable, principled way of life in System 4. This leaves System 5 optimistic about humanity, both individually and collectively.

If anything, System 6 carries that optimism even further. System 6 has the convenience of making its debut only after System 4 and System 5 have made sweeping social renovations. To the rule of principle and justice in System 4, System 5 has added widespread prosperity and abundant opportunities for individual achievement. System 6 therefore sees no reason to be pessimistic about humanity's prospects. As a matter of fact, System 6 is so confident about human nature that it may have little use for many of the constraints imposed by System 4. System 6 often views them as outmoded and unnecessary.

We should not leave the impression, however, that System 4 per se is a pessimistic modality. Overall it is highly optimistic. It derives that optimism from its confidence that ultimate truth will prevail. For the Christian this means an abiding trust that God's kingdom will indeed triumph over all that is vain and unrighteous. It is only when it comes to human nature that System 4 is pessimistic. Indeed, one reason it must trust so strongly in God is that it sees so little reason to trust in flesh and blood.

Suspicion and Fear

In a word, then, System 4 believes evil, not good lurks in the human heart. System 4 seriously doubts that human beings, left to their own devices, will truly pursue what is noble and right. It therefore views the human race with misgivings and suspicion, particularly those who are not part of its own inner circle. This basic distrust is why System 4 likes things black and white, why it operates on the principle, "Those who are not for us are against us." In System 4's judgment, if you are not helping promote its cause, you are working to overthrow that cause.

There is thus considerable insecurity and fear in System 4. This is understandable, given the precarious existence that System 4 maintained for centuries. In its struggle to topple System 3 in social and political affairs, constant vigilance was the price of survival. System 4 could never afford to drop its guard, even for a minute.

Systems 5 and 6, by contrast, while having known opposition, have never felt their very existence threatened to the degree that System 4 has. In the United States, Japan, and the nations of Western Europe, Systems 5 and 6 became widespread only after System 4 had secured general protection of individual liberties and freedoms. These later modalities developed in an atmosphere where they could express themselves without fear that some other system might rise up and crush them.

As a result, Systems 5 and 6 are non-plussed when they bump into distrust and fear in System 4. They cannot relate to that way of thinking. To them System 4 anxieties are unfounded, even irrational. Systems 5 and 6 are especially taken aback when they pursue a course of action with honest intent, only to discover that System 4 is suspect of their motives. System 5 and System 6 view themselves as alternatives to System 4, not a threat to it. So why does System 4 find it so disquieting to work with them?

Three Key Issues

We could answer that question by examining several contributing factors. But in the interest of brevity we will note only three. First is the problem that faces any modality when it tries to understand a system more complex than itself. For example, System 4 concepts make no sense to people who function in System 3 until a modicum of System 4 begins to appear in their makeup. System 7 outlooks elude System 6 until the System 7 begins to activate. And System 4 thinkers face the same challenge with System 5 and System 6. Once the mind develops at least a band of System 5 and System 6, it will start to see that insights from those modalities have merit. But absent that band, System 4 will not identify with System 5 and System 6 preferences.

Second, when it comes to the life of the church, System 5 and System 6 have a particular penchant for offending System 4, primarily in the way they speak of its values. To put it mildly, Systems 5 and 6 are frequently condescending or patronizing toward System 4. They rarely talk about tradition without a sneer in their voice. And despite the great cultural and missionary achievements of the System 4 church, Systems 5 and 6 seldom praise its accomplishments. In fact, they do just the opposite. They are more likely to ridicule the church of the past than to honor it. They easily leave the impression (because they often believe it) that yesterday's church lacked spiritual depth. On balance Systems 5 and 6 show no more appreciation of System 4's spirituality than System 4 does of theirs. Narrow-mindedness is a two-way street.

Needless to say, System 4 loses any desire to work with System 5 and System 6 if they repeatedly deride its values. System 5 can be especially disparaging of System 4 spirituality. System 5 is inclined to think that "new" means "improved," and "improved" means "better." William Barclay told of a librarian years ago who organized her shelves so that volumes written before 1900 were virtually inaccessible. This was not to protect the aging works from needless wear and tear. It was because she considered anything prior to the twentieth century of little interest or value. That librarian would be right at home among the System 5 elements who want nothing but contemporary hymns in worship.

Attitudes toward Change

Third, a great gulf separates System 4 from later systems when it comes to the subject of change. Systems 5 and 6 are often insensitive to that difference in the way they promote new ideas and directions for the church. They needlessly alienate System 4, or even polarize it. System 4 reacts by becoming entrenched. Not recognizing their mistake (which we will expand on below), System 5 and System 6 sometimes conclude that System 4 opposes change altogether.

To stigmatize System 4 this way, however, is patently unfair. Centuries of human progress occurred under System 4 tutelage, including the major philosophical developments that led to modern thought. History bears witness, therefore, that System 4 is not foreclosed to change. But System 4 does oppose change for the sake of change. System 4 has waged a tireless struggle against the undisciplined, impulsive lifestyle of System 3. To avoid further impulsiveness, System 4 wants change that is orderly, well thought out, and based on a solid rationale.

Systems 5 and 6, by contrast, are more concerned with openness to new directions than with the details of change itself. Change is a constant for them. Their day-to-day province is topsy-turvy, a place where everything seems fluid. To survive they must go with the flow, adjusting quickly to whatever happens next. System 5 is always looking over its shoulder, measuring its pace against its competitors, aware that it must innovate or be left choking in their dust.

To that end System 5 tinkers endlessly, trying this, then trying that, until it finds a way to do things better. In a word, System 5 lives with an open-ended commitment to change without knowing where change will take it. System 6 is similarly open-ended about the future. Relationships are as vital to System 6 as achievement is for System 5. But System 6 lets those relationships go where they may, with no preconditions as to the commitments or organizational initiatives that might ensue.

This means that Systems 5 and 6 have learned not to distract themselves worrying about the exact pattern change will take. In their world they rarely have the convenience of knowing how change will occur. They only know that it will. They have learned to cope with this uncertainty and not be preoccupied with it. For System 4, on the other hand, the framework of change is a primary concern. Before System 4 embraces a change process, it needs to know precisely what lies ahead. System 4 is never comfortable with change unless it knows what the limits are, what remains non-negotiable, and who will control the process. In short, System 4 needs a detailed blueprint of what the change entails.

Casting Things in Concrete

Unfortunately, Systems 5 and 6 are not well practiced in providing that blueprint. It is a skill they have few opportunities to use, since theirs is a world in which change is more commonly thrust upon them than intentionally pursued. They plan on the run. Even when they make intentional, strategic change, they have learned that their plans must be flexible and subject to constant revision. They are quick to assure their colleagues that "nothing is cast in concrete." Thus, when they talk about "new directions for the church," they can be long on generalities, but short on specifics. And that is where they lose System 4's support. System 4 *needs* things cast in concrete. It must feel assured that we are not opening the door for "anything goes."

Denied that assurance, System 4 digs in its heels. First it asks pointed questions, pressing for specifics about how these "new directions" would be undertaken. Because System 5 and System 6 are not accustomed to providing the level of detail that will help System 4 feel secure, their response may fail to soothe System 4 anxieties.

To make matters worse, Systems 5 and 6 may get a little testy over System 4's interrogation, for they will take it as a sign they are not trusted. But to System 4's way of thinking, change that is poorly thought out sounds like change for the sake of change. Afraid things are getting out of control, it sees no choice but to fight the System 5/System 6 agenda.

Allaying Fear and Anxiety

Once this occurs, we must quickly move to allay System 4's fears. Otherwise, its anxieties will grow unabated. As fear looms larger, System 4 may prepare for the worst by presuming the worst. A recent incident illustrates this point. Following a Sunday service that departed from our normal style, one member was quite upset. He talked about how much the change disturbed him, then asked, "Do we even believe in baptism any more?"

Now, Systems 5 and 6 find that reaction difficult to fathom. "Why should anyone think we are abandoning baptism," they ask, "just because we change the style of worship?" But if we put ourselves inside the System 4 thinking system, this reaction is altogether understandable. Because it is pessimistic about human nature, System 4 preaches respect for authority and insists on social structure and control. It easily presumes that people will throw off principle and restraint at the first opportunity. Due to this distrust, System 4 may not give Systems 5 and 6 the benefit of the doubt when they start nudging the norms. "What they are up to?" it wonders to itself. "Just how many standards do they intend to discard? All of them? That may well be."

Before we paint an unfair picture of System 4, however, we must add a qualifier. This reaction about baptism was admittedly somewhat extreme. It is by no means typical of System 4. On the other hand, neither is it altogether atypical. The deeper System 4's pessimism about human nature, the more likely an extreme response. System 4 will often warn about "throwing out the baby with the bath water" or "opening the floodgates." Another phrase System 4 loves is, "Give them an inch, and they'll take a mile." Whenever we hear such statements, we know that System 4 feels threatened.

And what is it afraid of? Mostly it fears a reversal of its hard-won victory over System 3. It does not trust System 3. System 3 is adroit at exploitation, and System 4 knows it. System 4 also knows that System 3 is shrewd enough to exploit any modality, including Systems 5 and 6. Consequently, System 4 maintains a constant vigil for System 3 passing itself off in System 5 or System 6 guise. When Systems 5 and 6 talk change, but leave specifics ill defined, System 4 becomes immediately suspicious. To System 4 this sounds like "anything goes," and "anything goes" means that System 3 has regained a foothold. Systems 5 and 6 may think they have proposed a change that marks a step forward. But System 4 may see it as a step back toward System 3.

Respect for Scripture

If it feels threatened enough by System 5/System 6 initiatives, System 4 is quick to take the offensive. (It learned long ago, in dealing with System 3, that he who hesitates perishes.) One common stratagem is to attack Systems 5 and 6 for not respecting Scripture. This is not necessarily a trumped-up charge. System 4 may well believe it is true. Our historical conditioning has created such strong mental models of how to "do church" — System 4 models, as we saw in chapter fifteen — that other approaches strike System 4 as unbiblical. Where System 4 has not learned to hold that impulse at bay, it will view any departure from System 4 methods as blatant disregard for biblical authority.

To charge Systems 5 and 6 with disregard for Scripture, however, is frequently unfair. Just because they offer alternatives to System 4 methodology does not mean they have dismissed the import of Scripture. Nor is it necessary to presume that respect for Biblical authority wanes just because System 5 or System 6 becomes dominant. To begin with, our dominant system has little to do with the value we attach to New Testament principles. The key is whether our dominant modality has an undergirding System 4 strata and whether our System 4 strata includes a high regard for God's Word.

These preconditions are necessary because we derive the doctrine of biblical authority from System 4. Without that modality our mind never raises the question of eternal or absolute truth.

Before we can honor biblical authority, therefore, our systems mix must include a vibrant System 4. Having an abiding respect for God's Word requires more than a hearty System 4, however. Many System 4 thinkers do not take their quest for truth to God's Word. Some turn to philosophy, others to natural science as their source of absolutes.

In fact, true philosophical atheism is as much a System 4 product as is systematic theology. System 4, like all the modalities, is a way of thinking, not a specific set of beliefs. We cannot anticipate the specifics of what people will think just because we know they conceptualize in System 4 categories. To take a parallel from another modality, System 2 societies are commonly polytheistic. Yet Abraham, himself a System 2 thinker in the midst of a System 2 culture, held firmly to belief in one God.

Thus, the first requirement for respecting biblical authority is an internal makeup that includes a well-nurtured, but not necessarily dominant System 4. Then, among our System 4 outlooks must be a conviction that life is answerable to biblical standards. If we develop that conviction in System 4, it does not go away simply because System 5 or System 6 (or any other modality, for that matter) becomes dominant. Again, modalities determine the *way* we think, not *what* we think. Nothing requires us to abandon fundamental insights from previous modalities just because we move to a new thinking system.

For that reason, Christians who develop a high regard for the Bible in System 4 may carry that esteem into later systems. They become understandably riled when System 4 charges them with indifference toward God's Word. Indifference is not the issue. The issue is a different set of lenses. Once Christians start reading Scripture through the eyes of Systems 5 and 6, things stand out in the text that they missed entirely before. As a consequence, System 5 and System 6 come away from Scripture with revised models of how the church should function.

Countering System 4 Anxiety

To allay System 4 anxieties, Systems 5 and 6 must continually demonstrate that they appreciate God's Word, that they are well-versed in it and respect it. Indirectly and inadvertently they often leave the opposite impression. This happens especially in classes and sermons taught from the perspective of System 5 or System 6. Where a preacher or a teacher is highly dominant in either of these modalities, there is a tendency for lessons to mention few specific Scriptures, if any. On the other hand, there may be several references to contemporary writers in the fields of human behavior, personal development, or even popular fiction.

Interestingly enough, these quotations from literature may perfectly accord with ideas found in the Bible. But the teacher often leaves that connection with Scripture implicit, never making it explicit, or at least not regularly throughout the lesson. The result is that System 4 begins to wonder if the speaker has a higher regard for secular authority than for biblical authority. In our seminars and workshops around the country we regularly hear System 4 members say, "Our minister preaches from everything but the Bible." In their voice is a tone that implies, "He doesn't preach from the Bible because he doesn't take it seriously."

As a rule, however, we know these ministers well enough to know that they love the Word and spend hours inside it each week. But their System 5/System 6 dominance lets them forget how vital it is for System 4 to see Scripture on bold display as the starting point for what we believe. Part of systems sensitivity is recognizing modality-specific needs such as that and addressing them in the way we structure lessons.

Some Basic Guidelines

From what we have discussed in this chapter, several guidelines become apparent. First, systems harmony does not "just happen." Leaders must purposefully promote it. They must teach it, practice it, and model it. Most importantly, leadership must learn to maintain systems harmony in its own ranks. If leaders cannot transcend the systems differences among themselves, they will never be able to manage systems tension on the broader front of congregational life.

Second, there must be a continual teaching program that

builds awareness of the church as a body. Most Christians are familiar with that analogy, but when conflict occurs, their behavior indicates that they have never internalized the values of the "one body" concept. We need to go beyond merely saying, "Paul teaches that the church is a body." People have heard that for generations, but it has not changed their conduct. Otherwise church splits would have disappeared years ago. Obviously something more is needed.

We would suggest that one missing ingredient has been straightforward, practical counsel on living out the implications of Paul's words. Periodically the pulpit needs to examine specific case studies in which people use biblical principles, especially the "one body" principle, to help them transcend their differences. Bible classes should do the same thing. Leaders have a key part in all of this. They must learn to react almost instinctually to conflict by asking the parties involved, "What are you doing to resolve this difference according to Biblical guidelines?"

Anticipating System 4's Needs

Third, we must learn to package proposals for change so that System 4 can feel comfortable with both the outcomes we desire and the process by which those outcomes are pursued. This means good, thorough planning in advance. It means getting the specifics worked out in detail. It means lots of one-on-one conversations with System 4 thinkers during the conceptual stage to find out what makes them uneasy with the new initiative under consideration.

We find it beneficial, indeed, to have two or three healthy System 4 thinkers on any critical task force that considers a major new work or ministry. These representatives of System 4 are there not only to offer the benefit of their personal insight, but to help the group understand the System 4 mindset in the congregation. When discussing a suggestion, those who chair a task force or committee often ask, "How would your friends in the church respond to this idea?" Because our friends normally have a systems mix similar to our own, the person who is System 4 dominant will typically answer this inquiry by describing the sensitivities of other System 4 people.

Such input is invaluable, for it allows us to consider modifications that would make our proposal less objectionable. It also allows us to anticipate criticisms and be prepared to deal with them constructively. As we move into formulating final plans, System 4 input can help us package and communicate our proposal so that it is likely to gain broad-based System 4 support. Sometimes we even ask System 4 representatives on the task force to be the ones who announce and advocate this initiative publicly.

Keeping Scripture in High Profile

Fourth, Systems 5 and 6 must learn to frame their recommendations with biblical metaphors, that is, with frequent allusions and references to Scripture and quotations of specific passages. System 4 must clearly see that the other modalities share its high regard for God's Word as ultimate authority. The reason System 5 and System 6 must make a conscientious effort to use Biblically flavored language is that they themselves are drawn to the language of the books they are most likely to buy. These include works on counseling, management, and self-improvement (in System 5), along with studies of therapy, emotional bonding, and personal transparency (in System 6). The language of that type of literature becomes a central feature of System 5/System 6 vocabulary.

System 4, however, does not read widely in those subject areas. It therefore raises its eyebrow when System 5 suggests (as we will discuss in the next chapter) that we need "an empowerment model of congregational management." To System 4 this sounds like System 5 is drawing its theology from the latest book it read on organizational theory. System 5 could make the same point, with far less likelihood of triggering System 4 suspicions, by saying, "We should structure the life of the congregation so that we are doing everything possible to equip Christians for effective ministry."

Notice the word "structure" that appears prominently at the first of this sentence. That term, in and of itself, is assuring to System 4. Then the proposed change itself is wrapped in the language of Ephesians 4:12, "equipping saints for the work of ministry." Once System 4 concurs that such an approach is both desirable and Biblical, fruitful dialogue can begin on how such a structure would function.

An Appropriate Vocabulary

As a footnote we might add that systems-sensitive management requires us to master a systems-sensitive vocabulary. Each modality responds warmly when it hears certain words and phrases. This subject is so important that we will devote a lengthy section to it later. For the moment we simply want to emphasize that the language we use has unparalleled impact on how systems respond to one another. We are not suggesting that we should manipulate people through language. But we are saying that we must learn which words to use, which to avoid in order to promote inter-systems understanding.

For example, we spoke earlier of teaching the congregation how to live together harmoniously, despite differences among us. We purposefully used the word "differences" rather than "diversity" in that case. We have found that many people who are System 4 dominant react negatively to the word "diversity." When they hear that word, they almost immediately think of *doctrinal* diversity. Needless to say, that gets System 4 dander's up. You may be surprised, therefore, that we put the word "diversity" in the subtitle of this book. The truth is, we deliberated for months before doing so. We knew that its prominence on the cover would cause many a red flag to go up in the mind of some would-be readers.

We obviously decided in the end to leave "diversity" in the subtitle, despite the risk involved. In the few words that fit on a book cover, we could find no better way to express what the reader would find inside than the phrase "empowering diversity without polarizing the church." But in congregational contexts the word "differences" is often a preferable term. It is far more neutral than "diversity" and not as likely to trigger unwanted reactions.

Despite the title of our book, where significant System 4

elements are part of a gathering, we talk about maintaining unity in the face of our differences, not in the face of our diversity. The principles are the same and the outcomes are the same, so there is no harm done (and great advantage gained) in using a term less likely to offend. Chapter twenty includes a lengthy system-by-system listing of terms that "connect" with specific modalities.

17

Diverse Expectations of the Church

In multi-system congregations, leaders must continually reconcile highly divergent convictions about how the church should function. Unfortunately, these conceptual differences are anything but superficial. They extend to the fundamental building blocks of congregational life, to such essentials as leadership roles, worship styles, ministry management, and Bible class structure. As systems multiply in a congregation, expectations in these arenas move farther and farther apart, until they sometimes seem all but contradictory.

A Class Exercise

Nor can we resolve the dilemma by asking which view is scriptural and which is not, for all sides defend their position with a biblical rationale. The problem is not disregard for New Testament principles, but disagreement over which biblical principles take priority. To illustrate, imagine a Bible class where the topic of the hour is Christian leadership. Systems 4, 5, and 6 are equally represented in the group. Now pose a question to them in this form:

> The New Testament describes several vital functions that leaders in the church should fulfill. Acts 20:28-31 says that elders are to be overseers of the congregation, protecting it from danger, false teaching, and internal disarray. Ephesians 4:11-12 portrays leaders as equipping saints for works of ministry. Then there is Peter's insistence that elders not lord it over the

flock, but serve instead as examples (1 Peter 5:2-3). So there are three differing roles, a managerial one (controlling what happens in the church), a training one (equipping members to serve), and a pastoral one (being examples to the flock). Now, leaders have a finite amount of time for their task. They must constantly set priorities. So if they cannot attend fully to all three of the duties we just outlined, which one should be their highest priority?

The conversation that ensues will be an intriguing display of systems divergence. Those who speak from the standpoint of System 4 will usually opt for the Acts 20 model. To them the first priority for leadership is to keep things organized, orderly, and under control. The System 5 representatives will tend toward the training model from Ephesians 4. From their perspective a church, no matter how well managed, fails in its mission if it is not empowering members to be optimally effective.

When System 6 takes the floor, a third position emerges. For System 6 the pastoral duties of 1 Peter should be leadership's overarching priority. System 6 takes this position because it perceives the shepherding process primarily as binding the emotional and spiritual wounds of hurting people. On the other hand, if we extend the definition of shepherding to include personal guidance and mentoring, System 5 may also select the 1 Peter model.

This little episode vividly demonstrates a point we have made from the outset. Namely, mental models change markedly as we move from one dominant modality to another. In the scenario we just examined, each conceptual system had its own leadership model, and each model drew on apostolic language. But the priorities in the System 4 model differed pointedly from the ones in System 5, which in turn departed sharply from those in System 6. How does a leader measure up to such contrasting demands from the congregation?

Then add another problem. These diverse expectations, demanding in their own right, are only the tip of the iceberg. Beneath the surface other models, equally divergent, set one system against another with regard to basic congregational missions.

A System 4/System 5 Contrast

To see how deep this divergence runs, we need look no further than Systems 4 and 5. Their respective priorities for the church are a study in contrasts. We see this, for instance, in the way they view the church as a teacher. Both agree that the church has a crucial training role.

But System 4 describes that role in terms of indoctrination, System 5 in terms of empowerment. Ask System 4 to identify the church's fundamental mandate as a body, and the answer will normally boil down to "upholding the cause of truth." Put the same question to System 5 and the reply will probably reduce to "helping people find an abundant life in Christ."

Thus, where System 4 is highly dominant, with other systems only marginally present, church life centers on teaching events. And the purpose for these events is almost exclusively instruction in God's Word. A sermon, usually brimming with biblical citations, dominates the worship hour. Classes are literally Bible classes, based on either straightforward study of Scripture or the examination of doctrinal issues.

There are also probably classes in the middle of the week, offering members a miniaturized version of what happens on Sunday mornings. In addition, a guest speaker comes once or twice a year to preach evangelistic lessons for several consecutive days. Indoctrination, in short, goes on everywhere. Even the weekly church bulletin is replete with "sermonettes."

When System 5 becomes dominant, it may retain the structural features of classes and instruction, but their flavor changes. For instance, outside speakers may still come in for several days, but they are more likely to talk about marriage enrichment than preach evangelistically to reach the lost. Bible classes start having titles like "How to Cope with Depression" or "Managing Your Time More Effectively." Even though the teachers may address these topics within the bounds of biblical guidelines, the tone of the Bible school program has clearly shifted.

Basic textual studies will still be part of a System 5 environment, for System 5 is eager to know God's Word. As compared to System 4, however, System 5 is not inclined toward studying truth for the sake of truth. System 5 zeroes in on truth that makes an immediate and worthwhile difference. Where System 4 will study Deuteronomy just because it is part of God's Word, System 5, sitting in the same class, will continually grouse, "What difference does this make? This is not helping me draw one bit closer to Christ or to be more effective in my daily walk!"

Mushrooming Knowledge

From the vantage of System 5, necessity all but dictates this pragmatic approach to study. By the time System 5 arrives on the scene, we are struggling madly against an avalanche of information. To make matters worse there are incessant demands to become familiar with ever-changing technology. We can hardly keep abreast of vital developments in our own professional arena, much less step beyond that arena and pursue the broad realm of essential knowledge.

By "essential knowledge" we mean the body of truth and technical know-how that a society must sustain in order to perpetuate its existence. Less than two centuries ago, at the height of System 4 ascendancy, it was still possible for a single individual to master almost all essential knowledge. Universal geniuses like Thomas Jefferson and Benjamin Franklin, while not found in every village, were hardly an oddity. Buoyed by their example, people applied themselves assiduously to truth. Ordinary citizens were eager to explore a subject just because it was true. After all, a lifetime was long enough to become a walking encyclopedia of most human learning. People would therefore sit patiently through three and four hour lectures on almost any topic imaginable.

Today no one could aspire to universal genius. Essential knowledge is simply exploding, doubling in scope every few years. As a consequence, we must daily decide what knowledge to pursue, what knowledge to forego. What we choose to forego, moreover, is a volume of truth whose magnitude far exceeds the knowledge we have time for. The critical information we do *not* know grows larger every day. Faced with that reality, System 5 no longer feels the luxury of pursuing "truth for truth's sake." It applies itself primarily to truth that has a near-term impact on enterprises at hand.

When System 5 goes to church, it maintains this same pragmatic outlook toward study. As opposed to System 4's emphasis on doctrinal theology, System 5 wants sermons that stress practical theology. System 5 loves the word "relevant." Classes and preaching should be "relevant," it insists. Relevance does not mean the elimination of doctrinal content, as System 5 is quick to point out. But to strike a resonant response in System 5, sermons must be packed with workable insights for personal effectiveness. Theory for theory's sake has little appeal to System 5. Nor does theology for theology's sake. System 5 is looking for truth that makes an immediate difference in the issues of day-to-day life.

An Empowerment Model

Moving beyond the classroom, System 5 takes its preoccupation with personal relevance everywhere it goes in the church. For that reason, System 5 gives priority to spiritual giftedness in a way that System 4 seldom does. System 4 emphasizes the implication of truth. System 5 stresses the implication of being gifted personally by God. To cite an example, System 4 might say, "Women should not lead public worship, and here are the passages that say so." System 5 might rejoinder, "God surely would not give my daughter a gift that He did not intend her to use. Therefore, we may be reading those passages amiss. Let's reexamine them."

In a word, System 5 views the church as less in the business of indoctrinating members and more in the business of equipping and empowering them. System 5 may agree that indoctrination is essential to empowerment, but it does not make doctrine the end-all of training. Alongside "head knowledge" System 5 insists on an educational program that enhances spiritual and personal skills. This is another way of saying that System 5 supplants a pedagogical model of the church (from System 4) with an empowerment model.

But when we start using an empowerment model in congregational management, we create an immediate impact on organizational structures. System 5 thinks it pointless to help people develop their gifts, then offer no forum in which to exercise them. Opportunities for personal, fulfilling ministry are vital to System 5, and it wants those experiences as much for women as for men.

It therefore expects leadership to create a multi-faceted array of ministries, so that every member of the Body finds a place of meaningful service. And the stress is on "meaningful." System 5 almost invariably lives on a tight schedule. Nothing irritates it like being asked to do something it views as busywork.

Taking Initiative

System 5 also wants individuals to be free to take broad initiatives in the ministries they choose. It thus promotes dispersed decision-making in a congregation. System 5 appreciates broad guidelines and policies from leadership, but within that guidance System 5 likes to set its own pace and direction. It keeps leaders informed, but does not feel duty bound to obtain their approval for everything it does.

This approach is a wholesale contrast to System 4, which entrusts decisions to a handful of people at most. Only by centralized decision-making, System 4 believes, can we "keep things under control." Thus, System 4 usually exercises minimal discretion as a follower. When given a job to do, it refers anything of significance to leadership for review and consent.

Commitment to an empowerment model also causes System 5 to differ with System 4 on staffing. Because System 4 views the church first and foremost as a teaching institution, System 4 sees no need to have more ministers on staff than preaching and educational requirements dictate. Yet, as System 5 gains influence, ministerial staffs inevitably grow larger. Since System 5 wants far-ranging programs of ministry, coordinating all that activity demands full-time attention. Administrative duties begin to shoulder out the time ministers once had for study and teaching. As a result, System 5 churches promote ministerial specialization. One or two staff members are charged with being "ministers of the Word." The rest take on managerial roles that may never place them in the pulpit.

Adding System 6 to the Mix

By now you are beginning to see the kind of challenge that leadership faces when Systems 4 and 5 sit shoulder to shoulder in the pew. In almost every essential area of work and worship these systems are disagreed on priorities and styles. So what happens when we add System 6 to the mix? What further complexities set in?

To begin with, everything gets smaller and slower if System 6 has its way. From a managerial standpoint, System 6 is almost the exact opposite of Systems 3 and 4. Where those modalities restrict decision-making to a tiny cadre of leaders, System 6 thinks everything should be done by consensus.

System 6 would point to the Jerusalem conference in Acts 15 as a model of how churches should operate. There the apostles and elders, along with the Jerusalem church, spent tireless hours to develop an agreement on circumcision. It makes little difference to System 6 that this collaborative approach required days of discussion and dialogue to resolve the issue at hand. The important thing is that those discussions ended in accord, having worked out a conciliatory approach to an emotional and thorny problem.

It is this pursuit of consensus that slows things down in System 6. To give every party an opportunity to say its piece, System 6 refuses to be hurried. To build true consensus, it must also work in groups that are intimate, where there is complete openness and no hidden agendas. This is true whether we observe System 6 at work on a committee, on a project of service, or in settings of worship and study.

Moreover, System 6 seeks an intimacy that is frequently possible only in very small groups. Otherwise there are too many people for each person to know the heart of everyone else. When System 6 groups get together, they typically tell each other lots of stories, particularly stories that are self-revealing, for that is how they come to truly bond with one another.

Putting System 4 and System 6 Together

System 6 is so different from System 4 in style that it is a challenge for them to work together on a committee or task force. The fundamental problem is that System 4 quickly becomes frustrated with all the story-telling and consensus-building in System 6.

"They never get anything done!" System 4 complains in its frustration. "They just sit around and talk." When System 4 is placed in charge of a project, it moves to structural issues immediately. Within one or two meetings it has put together its organization, outlined specific objectives, and set up a system of deadlines and accountability.

By comparison, if System 6 is organizing a project, the formative meetings look loose-ended, as though they are going nowhere for weeks on end. But these seemingly non-structured sessions are in fact quite essential. System 6 uses them to build interpersonal understanding within the group so that the members can reach a solid consensus when the time for tough decisions is at hand.

This casual, story-telling, life-sharing atmosphere that System 6 prefers carries over into its expectations for worship. It seeks a service that is "touchy-feely," to borrow a term other modalities often use to describe what they see in System 6. Along with that "touchy-feely" atmosphere, System 6 wants sermons that are heavy with stories and narrative content. System 6 points to the parables of Jesus as indicative of what preaching should be.

Narrative teaching, modeled on the style of Jesus, connects far more readily with System 6 than does a doctrinal discourse based on the style of Paul's letters. There is an intriguing tendency, indeed, for System 4 to preach and teach more comfortably from the Epistles than from the Gospels, while in System 6 the Gospels hold more attraction than do the Epistles.

Cutting Out Administration

Organizationally, too, System 6 takes a different tack from previous systems. To put it mildly, System 6 wants as little organization as possible. The church should be in the business of ministering to people, System 6 holds, not meeting to take care of business. One person summed up the System 6 view succinctly when he said, "Church is a place to tend to people, not a place for people to attend." System 6 is hardly so naive as to believe that we can dispense with organization altogether. But it strives to keep structure to a minimum. In System 6's view the church is so weighed down with organization, decision-making, and administrative matters that it has no time to care for people.

To reduce the administrative overhead in the church, System 6 would have as few rules and policies as possible. System 6 is often an outspoken critic of tradition, for its sees tradition as merely another set of needless expectations to meet. System 6 believes that leadership should focus on broad guidance and vision-casting, but entrust decision-making to ministry coordinators and front-line care providers. Since they are closest to the need, they know what best to do.

The Need to Consult

Nevertheless, whatever the level at which decisions are made, System 6 expects to be consulted. So does System 5. In the transition from System 4 to Systems 5 and 6, leaders must learn to include far more people in the decision-making process. Leading a System 4 organization does not condition us to create broad involvement in decision-making. System 4 followers, after all, are more likely to wait for guidance than to demand a voice in policy. Because it prides itself on being trustworthy and devoted to duty, System 4 typically supports what leaders decide, whether decision-makers consult the rank and file or not.

Systems 5 and 6, on the other hand, want to be heard during the deliberative stage. Neither of them insists on having its own way. Like System 4, they can support leadership initiatives that do not fully align with their preferences. But System 5 and 6 want to know that their viewpoint was heard, understood, and taken into consideration in the final decision.

Parenthetically, we might add, we are seeing more and more signs that System 4 wants to be consulted, too. So long as congregations enlisted their leadership primarily from System 4 thinkers, members who were themselves System 4 dominant had no doubt that their preferences were known in leadership circles. Today, as congregations increasingly draw their leaders from System 5 and System 6 circles, System 4 no longer has that assurance. It therefore is beginning to ask more probing questions as to how certain decisions were reached and what factors shaped the thinking of those who made them.

18

Four Essentials
in a Multi-System Church

The diversity we outlined in chapter seventeen is what causes leaders to ask at times, "Can we do anything right?" No matter what they decide, someone seems certain to object. Is there any way to blend the expectation of Systems 4, 5, and 6 into a harmonious, working whole? We believe there is. But it requires leadership to practice a whole new set of organizational skills. For the most part, churches have long enjoyed the luxury of "one-size-fits-all" programming. Each Bible class tended to look like a clone of the one next door. Every worship service had a near-identical structure to the one the week before. All committees and ministries were tasked and managed in exactly the same way.

In a setting like that, the function of leadership was much less complicated than it is in a multi-system church. Bible classes designed to meet System 4 expectations will not look like those aimed at meeting System 6 needs. Neither will worship services. We will not manage System 5 volunteers the same way we direct those who are System 4 dominant. In essence, we must tailor-make every structural element in a multi-system church. Doing so requires immense leadership energy, far beyond what traditional church organization demands. Not only that, multi-system management puts a premium on planning carefully, gathering feedback continuously, and making frequent midcourse corrections. You cannot put a multi-system church on autopilot and expect it to fly straight and level very long.

How, then, should we proceed? Over the next few pages we

explore four essential duties that are incumbent on leadership in a multi-system church. These four responsibilities are the building blocks, so to speak, on which the entire enterprise of systems-sensitive leadership stands. Without intending to imply a priority, we look at them in this order.

- Developing a congregation-wide atmosphere of forbearance
- Maintaining feedback loops in decision-making
- Practicing systems-sensitivity without using systems terminology
- Becoming diligent in vision-casting

Another Look at Forbearance

This is actually our second time to discuss forbearance as an essential. We also mentioned it in chapter fifteen. There we talked about the need to build a love for harmony and unity throughout the congregation. Now we return to that theme because it is so critical. Even when all goes well, there are enough differences between Systems 4, 5, and 6 to make friction an ever-present danger. If those modalities do not learn to approach each other consistently with a forbearing spirit, conflict is sure to have its day.

For a multi-system church to succeed, we must regularly and consistently hold up the ideal of unity. And we must do so in countless forums, not just occasional ones. In effect we are trying to create a tradition that says, "We don't fight here. We find more responsible ways to handle our differences."

Building a Heritage of Peace

To build that heritage, leaders must first effect systems harmony in their own ranks. Rarely will an organization rise above the performance of its leadership. If leaders cannot transcend systems differences among themselves, the congregation is unlikely to fare much better. As for the congregation, our goal is to foster a climate in which forbearance prevails. This forbearance expresses itself in a universal feeling that:

- Change to accommodate diverse needs is regular and routine around here and nothing to be alarmed about.

- Those changes may require me to forego my own preferences at times. But we all defer to one another in this church.
- In deferring to others I can be confident that my own preferences are still respected. I also know that they will be honored on other occasions.

We should strive to reinforce such feelings whenever we make significant changes. Some typical moments at which we might need to reemphasize forbearance would be:
- revising the worship format, even for a single service
- introducing a new hymnal
- rearranging the seating for worship
- restructuring Bible classes
- moving the activity of a major ministry to a new location
- modifying the schedule for congregation-wide events
- making facility modifications
- launching a new ministry that will require significant congregational resources
- discontinuing an event or ministry that has been part of the congregational heritage for years

These types of initiatives are commonly a source of dissension or grumbling. To offset that prospect, we should never undertake them without reminding the church directly or indirectly:

- that Christ paid a dear price for our oneness
- that we dishonor Him when we let division get a toehold
- that to avoid division we must defer to one another
- that deference should be a two-way street, so that no one set of preferences holds constant sway, with everyone else always required to conform

Our reminder need not be elaborate or involved. It might consist of nothing more than a statement like this: "Today's service includes a lengthy reading of Matthew 26 while the communion is passed. We know that many of you find the Lord's Supper more meaningful if we have silence during that period. And we will do it that way again next time. But today we are going to be sensitive to our brothers and sisters who gain more from the communion if a Scripture reading keeps their mind focused on Christ's passion."

This statement embodies several elements that should always be present when we propose a significant change.

- It identifies a specific congregational need that makes this change appropriate.

- It outlines precisely what the change will be.

- It shows sensitivity to those whose personal preferences do not accord with the change.

- It solicits their willingness to defer to the needs of others.

- It assures them that we are not overlooking their own preferences and that we will honor those preferences, too.

Considerations in Appealing for Unity

You will notice, however, that this statement included no direct appeal for unity. Nor did it emphasize how much Jesus cherishes our harmony. Should we leave that appeal implicit? Or should we be explicit about it?

There is no rigid answer to that question. Clearly there are times when we *should* be forceful in calling for unity. But there is no need to follow that practice every time we announce a change. Doing so, indeed, runs a danger. It invites a kind of speculation that may actually trigger dissension, not minimize it. Some people will begin to say to themselves, "Why do our leaders feel compelled to talk about unity every time they announce something new? They must be afraid that these changes are potentially quite divisive. If that's the case, then maybe we ought to forego them."

There are times, therefore, when we want to remind the congregation of the unity mandate without referring to it directly. To make that possible, we must build a strong link in every mind between the price Christ paid for unity and our need to defer to one another. We create that link from the pulpit, in Bible classes, by remarks made in planning sessions, through bulletin articles, with congregational slogans — in short, using every avenue of communication.

Once we establish that link, and establish it firmly, we can often be indirect in our appeal for unity. We do not have to be explicit every time. We will be able to talk about deferring to one another, never mentioning oneness in Christ. Yet the congregation will recall the mandate for unity automatically.

Nevertheless, even with such linkage in place, we should not rely on indirect appeals entirely. There should still be occasions when announcements of change call forthrightly for harmony and forbearance. For one thing this helps new members experience the congregation's commitment to diversity. And second, it takes the subject of forbearance out of the theoretical and gives people a specific instance in which to apply it.

When we do choose to be explicit, one other guideline applies. We should not begin an announcement of change with an appeal for unity. We should place that appeal at the end. If we start off by underscoring the need for unity, we again invite counterproductive speculation. People will say to themselves, "Why all these opening remarks about unity? Our leaders are obviously about to announce something. Whatever it is, they must be worried that it could really cause us trouble." Once people start thinking that way, even a routine change may leave them unsettled. "Well, on the surface I don't see anything dangerous about this idea," they may muse to themselves. "But maybe there is something I've missed."

To offset that prospect, not only should we place the unity appeal toward the end. We should often treat it almost like an afterthought, as something that really did not need to be said, but we will say it anyway. Neither does it have to be in the form of an appeal per se. It might be nothing more than a concluding statement that says, "We appreciate the loving way you always accept a change like this. How great to be part of a church that is so gracious and selfless when asked to defer to someone else's need. Your kindness and thoughtfulness toward one another are the very thing Jesus had on His heart as He prayed for unity in His final hours."

Feedback Loops

To know where change is needed and to manage it effectively, leaders must have the pulse of all the systems in their congregation. Equally important, leaders need to know how well previous change has been received or is perceived as working. To those ends a second essential for systems-sensitive management is a purposeful program to gather continual feedback.

Commonly this means wide use of surveys. In a multi-system church, surveys should become as much the way of doing business as an opening prayer at Sunday services. We are not talking here about huge, lengthy questionnaires. For one thing, they are difficult to design well. They also require so much time to administer that working them into congregational schedules is a challenge. In addition, it is difficult to justify one of these exhaustive surveys more often than every two or three years. But in today's rapidly changing world, survey data that is two or three years old has limited value.

What we propose are less cumbersome feedback loops. By sampling small segments of the congregation on limited topics, but doing so regularly, we get valuable information in a timely, non-disruptive fashion. Periodic focus groups are one way to accomplish this, as are congregational open forums on specific issues. Brief questionnaires, perhaps asking for a half-dozen responses, are another.

One congregation puts twenty inserts at random in its program of worship, asking the person who receives that copy of the program to evaluate the morning service. The survey, in the form of a questionnaire, can be completed in just a few moments. It is then left on the pew to be collected for review. Because of the random distribution, worship planners get immediate feedback from differing viewpoints (and differing systems) each week. Over the course of three or four months leaders are able to sample most of the congregation, with valuable insights gained weekly. Not only that, by completing the questionnaire while the day's service is a fresh memory, worshipers often comment on items that might not come to mind in a massive survey months later.

Don't Leave Feedback to Chance

Without this type of purposeful effort to uncover congregational sentiments, we must rely on intuition or chance conversation to determine how people feel. But word-of-mouth feedback is risky, at best. When System 4 dislikes an initiative, leaders know it immediately. On the other hand, those same leaders may hear nothing from System 5 or System 6. Absent that feedback, leaders often conclude that System 4 speaks for the entire congregation. That sets the stage, then, for reversing course and abandoning an apparently troublesome change.

Once they make an "about face," however, it is not uncommon for leaders to hear from Systems 5 and 6 en masse! It turns out that these systems had been thrilled with the now canceled initiative. What the leaders thus have on their hands is a situation in which all parties are upset. System 4 cannot understand why the church took such an ill-advised initiative to begin with. And Systems 5 and 6 are perturbed that leaders backed away from a promising direction just because System 4 complained.

Had leadership practiced more astute systems management, they might have avoided this predicament. Systems-sensitive leaders never leave feedback to chance. They continually look for opportunities to tap into congregational thinking. They periodically think through the congregational issues that have a high potential for creating systems tension. They identify the questions they need to answer in order to be certain of the attitudes surrounding those issues.

Then they find an appropriate forum or mechanism for gaining answers to their questions. They often take advantage of small groups that already exist — Bible classes, support groups, LIFE groups, ministry teams, etc. — by using their meetings as occasional information gathering opportunities, identifying what individuals in the group think about a given issue. In short, planning to obtain feedback and planning how to do it are as central to the way leadership functions as developing an annual budget.

Tell Them What You Learned

As we indicated above, surveys will always be a prominent part of a feedback system. Be aware, however, that conceptual systems will react to surveys in significantly different ways. For instance, Systems 5 and 6 will be more enthusiastic about them than System

4. Because of its sense of duty, System 4 will respond to a survey simply because leadership asked it to. Systems 5 and 6, on the other hand, are thrilled when leadership requests their opinion. System 5 and System 6 want to be consulted.

But there is a catch here. They also want to know that something worthwhile is happening with the survey data. They quickly sour on the prospect of filling out forms that serve little practical purpose. This happens with System 5 because its schedule, always tightly packed, allows little time for busywork. And with System 6 a form for the sake of a form exemplifies the very thing it disdains in organizational life, namely, more focus on perpetuating the organization than on caring for people.

When asked to complete a survey, therefore, Systems 5 and 6 want to know why this information is important and what will be done with it. We do not have to answer those questions every time we present a questionnaire. But leaders must nurture an atmosphere in which people feel that their input makes a difference. To do that, leadership must regularly explain how survey responses have shaped specific decisions. Leaders must also make timely reports (in days, not months) following major surveys, summarizing enough of the findings to give the church a feel for what was learned. This is leadership's part in completing the feedback loop that the congregation's input began.

A Lesson Learned the Hard Way

Several years ago we alienated a talented System 5 professional by overlooking this aspect of the feedback principle. He had put together a massive and masterful marketing study for the church. He provided a detailed report on his findings. Over the next 18 months, in one leadership meeting after another, we went back to the results of his work. His research colored a host of major decisions and provided the basis for a two-day planning retreat. But no one ever communicated that to him.

We did not recognize our mistake until we received a letter from him, informing us that he was transferring his membership to another church. In the letter he described his disappointment that all his work on the marketing study had meant so little. As far as he was concerned, his report and recommendations had been for naught, relegated to some long-forgotten shelf to gather dust. We learned from that experience never to presume that people see the connection between their own input and the decisions that come from leadership. We must be specific in the matter. We must complete the feedback loop.

When System 5 and System 6 ask for feedback from leadership, they are not demanding that their views prevail. They are willing to rally around a decision that goes against their preferences. But they must know that decision-makers took their view into account. This feedback need not be extensive. It can often be accomplished in three or four sentences: "We appreciate your thoughtful response to our questionnaire about the date and format for VBS. You offered a number of excellent suggestions. On balance, however, it appears that parents prefer the same time frame and format we used last summer. Before we schedule next year's VBS, we will ask for your input once more to see if preferences have changed."

Incidentally, leadership feedback to the congregation is important whether input from the church results in change or not. Leaders sometimes mistakenly conclude that if a study of congregational opinions supports maintaining the present course, there is no need to report findings back to the church. The purpose of feedback is not just to report results, however. It is also intended to assure people that their voice was heard. And they need that assurance regardless of survey outcomes.

Do It, Don't Talk It

A third essential in multi-system congregations is to be systems-sensitive in your leadership without using systems jargon in congregational communication. You will notice, for instance, that our examples of communicating with a congregation, both in this chapter and throughout the book, avoid systems terminology. We encourage you to follow a similar pattern yourself.

We offer that counsel from our own experience. If you find systems thinking as helpful as we have, you will be tempted to start "talking systems" everywhere. But proceed carefully. Unless *everyone* in your congregation understands the modalities, and understands them well, you will inhibit communication when you speak of "System 6 values" or "System 4 priorities." Many people will have no idea what you are talking about. Either they will immediately feel like an outsider, or else they will write you off as some type of elitist. In neither case have you engendered a constructive reaction.

We could overcome that problem, it might seem, if we simply introduced the entire congregation to the conceptual systems. Unfortunately, at least three factors make that solution unworkable. First, not everyone sees the value of applying systems insights to the church. Systems approaches to management are a product of System 7 outlooks. Unless people already have at least an emerging band of System 7, they will react to "systems talk" as irrelevant or even boring.

Second, there is the matter of time. People do not hurriedly become conversant with the various modalities. While we can sketch the basics of the eight systems in just a few minutes, truly comprehending them requires hours of work. Look at how many chapters we took to give you a fundamental sense of the individual systems. Where will you ever find the setting to offer such extensive training to every member of your congregation? And what about new people who are coming into your fellowship all the time? How will you quickly induct them into this type of systems thinking?

Third, in the eyes of many (especially System 4), leadership immediately discredits itself if it leaves the impression that sociological or psychological theory is setting the agenda instead of Scripture. In churches with a high regard for biblical authority, leaders must justify their actions consistently from the Word. Fortunately, systems sensitivity allows us to do that very thing. The Bible is replete with systems themes, even though it never uses systems terminology. The rationale for almost every systems initiative, therefore, can be taken from Scripture without turning to psychological jargon.

In summary, we need to reserve "systems talk" to private conversations and to planning sessions among leaders. Put your emphasis on teaching systems-sensitivity to leaders, not the entire congregation. You can "capture" the entire leadership for training more easily than you can the total membership. And in today's church every leader needs systems skills.

Vision-Casting

The fourth essential in a multi-system church is for leaders to take vision-casting seriously. Almost everyone acknowledges that leadership has a duty to articulate vision and purpose. But acknowledging that importance is one thing. Believing in it passionately is another. Historically church leaders have been able to get by with little more than lip service to vision-casting. As long as members were coming dominantly from a System 4 outlook, there was almost a natural consensus about what the church should be. That consensus served as a surrogate uniting vision, even if leaders were not particularly astute at defining and stating direction. In addition, System 4 followers were often compliant enough to muster behind leadership, even if congregational leaders had no vision to guide their choice of direction.

Once other systems become prominent in the church, this natural consensus starts to wane. Today, if leaders fail in the role of vision-casting, nothing else can fill the void.

- First, leaders must have a clear picture in their own minds as to where the church is going.
- Second, they must identify the key strategies that will move the church from where it is to where it needs to be.
- Third, that vision and those strategies must become internalized in each leader's heart. They should come to mind immediately and succinctly every time a leader faces a consequential decision or is asked about congregational plans.
- Fourth, leaders must work tirelessly to find simple, clear-cut ways in which to explain this vision and the strategies that support it.
- And fifth, leadership must communicate, communicate, communicate to the congregation.

The initial target audience for this communication is congregational "middle management," the people who are doing hands-on ministry planning. They should be able to articulate the vision, along with the strategies to achieve it, as instinctively as those with overall congregational responsibilities. Otherwise, the secondary level of leadership will start laying ministry plans that inadequately support — or even work against — the congregational vision.

The second target for leadership communication is the congregation itself. Almost without exception, leaders presume that the man and woman in the pew understand the church's vision and direction better than they actually do. When people start complaining that leaders have talked too much about vision and goals, then maybe they have. Until that time, we should presume we have not said enough, not nearly enough, about where we are going and what our principal priorities must be for the foreseeable future. People will not remain focused on the vision unless leaders remind them of it over and over again.

Committing to Diversity

A key element of that vision must be a commitment to honor and build on diversity. If leaders are not willing to go on record as believing that diversity is a strength, not a threat, there is no point in attempting systems-sensitive management. Your only recourse (other than continued friction and conflict) is to start planting daughter churches that build around the systems you are unable to accommodate inside your own walls.

The problem with this approach is that it only postpones the inevitable. People do not remain "frozen" in their present dominant modalities. It will only be a matter of time before both the mother church and the daughter church are struggling with new emerging systems in their midst. The wiser approach, therefore, is to harness diversity now and structure our vision around it.

Many excellent books, including most of the best-selling works on contemporary organizational management, devote lengthy sections to the issue of how to build and sustain vision.[1] We see no

---

[1] Some of the finer materials on this topic will be found in Warren Bennis and Burt Nanus, *Leaders: The Strategies for Taking Charge* (New York: Harper & Row, 1985), pp. 87-109; Peter Koestenbaum, *Leadership: The Inner Side*

point in repeating that material here. We do think it necessary, however, to underscore that times have changed and church leaders can no longer default in their vision-casting role. Otherwise, as they move toward an empowerment model for the church (which they *must* do if they intend to hold System 5), they will actually infuse their congregation with confusion and conflict. When we empower people, but do not provide a uniting vision, we invite them to pursue their own individual visions. Once we do that, we set in motion an inevitable clash of individual visions and the prospect of key ministries working at cross-purposes.

So long as decision-making was centralized and held in the hands of a small leadership core (the historic System 4 pattern), it did not matter that leaders were the only ones with a shared vision. Since they made most of the decisions, ministry activities tended to support a rather common sense of direction. Because an empowerment model, by its very nature, disperses decision-making, the absence of a uniting vision invites every ministry to go its own way. Like Israel in the days of the judges, everyone does what is right in his or her own sight (Judges 21:25). The result was chaos and anarchy in Israel. The same thing will happen in the church.

of Greatness (San Francisco: Jossey-Bass Publishers, 1991), pp. 105-135; James M. Kouzes and Barry Z. Posner, *The Leadership Challenge: How to Get Extraordinary Things Done in Organizations* (San Francisco: Jossey-Bass Publishers, 1991), pp. 79-130; Peter M. Senge, *The Fifth Discipline: The Art & Practice of the Learning Organization* (New York: Doubleday, 1990), pp. 205-232.

19

Systems-Sensitive Bible Classes

B eyond vision-casting, leaders in a multi-system congrega- tion must program toward the very diversity they have pledged to honor. One place this occurs is in Bible school classes. In the typical congregational setting, most classes are age-defined: young singles, young couples, mid-adults, senior adults, etc. That structure has served well for generations. But the day has come when we need to consider another. To be specific, we need to develop educational strategies that build around systems common- alities. Since age is no determinant of dominant modalities, we have young people and old alike who rely primarily on System 4 thinking. We have others, young and old, who are System 5 thinkers. Then still another segment is System 6 dominant.

These modalities have altogether different preferences when it comes to class styles and methodology. But so long as we maintain age-defined classes, all the 40-year olds are in the same setting, no matter what their dominant system is. Which system shall the class cater to? No matter which one we choose, we do not optimize the learning experience for the other systems. Not only that, we encourage disinterest and dropout among those whose system needs are neglected.

Systems 4 And 5 in the Classroom

Our traditional Bible class model was bequeathed to the church by System 4. As we might expect, given System 4's concern with doctrine, this model presumes that classes should primarily aim at indoctrination. The structure is content-oriented. That is, the more content communicated in an hour the better. In a System 4 Bible class the overriding qualification for a teacher is how well he or she knows Scripture and doctrine. Good teaching techniques, while important, are clearly secondary. Students are to walk out of a System 4 Bible class knowing the Word in depth, knowing what they believe, and knowing how to defend it in any setting.

This type of class, while certainly having a vital place, does not hold the attention of System 5. When System 5 turns to the church, it is not looking first and foremost for indoctrination. It wants a strong doctrinal base, to be sure. But it primarily seeks help in a discovery process. When it opens the Bible, it wants guidance on how to discover for itself the truth contained there. It is not particularly interested in being told what to believe by someone else, no matter how knowledgeable and authoritative that person may be. In its encounter with Scripture, System 5 hopes to uncover new vistas of self-understanding and personal effectiveness.

A Discovery Process

Moreover, we talk about a "discovery process" with System 5 for good reason. System 5 learns through hands-on experiences — either real-life encounters or carefully crafted simulations — that offer moments of breakthrough learning. As a result, System 5 loves classes that build creatively on role playing, drama, and case studies. These methods also appeal to another characteristic of System 5, namely that it learns by watching a story unfold.

This stands in contrast to System 4, which learns by hearing the facts laid out. System 4 typically views role playing and case studies as a waste of precious class time. It believes the proper way to conduct a class is for a recognized authority to share knowledge informatively with a largely passive audience. The audience is present to hear, not to interact.

System 4 can be quite satisfied, as a result, in huge classes or in lecture hall settings, listening to a presentation that goes unpunctuated with visuals, graphics, or group discussion. If visual aids are used, System 4 will benefit from them, to be sure. But it will not leave a class feeling unstimulated just because they were omitted from the lesson plan.

This penchant for lecture is why the sermon grew to dominate the worship hour in System 4 congregations. People who are System 4 dominant normally prove to be superior aural learners. Contrast that to System 5, which tends to learn visually. When a lesson receives no visual reinforcement, System 5 *does* feel that something is amiss. For one thing, visual representation can simplify concepts, relationships, and information in a way that oral description cannot. System 5 comes to prominence in a world that is both complex and overloaded with data. It depends on visual presentations to help it get to the essence of the subject at hand.

Beyond that, System 5 is simply not the aural learner that System 4 is. System 5 needs visual input to grasp a subject thoroughly. This is not to say that the aural message is unimportant to System 5. To the contrary, it is most important. But aural input needs the catalyst of visualization to "get through" to the System 5 mind.

In a sense, System 5 learns the same way that all of us respond to television. Even though we think of TV as a visual medium, its message content is primarily aural. If you stand in an adjoining room from a TV set, where you hear the sound but cannot see the picture, you can probably follow 95% of what is happening.

But have you ever tried to watch television with the sound turned off? Apart from sporting events, some cartoons, and a few commercials, video alone becomes quickly meaningless. Without the sound track you cannot make sense of what occurs. Why, then, do we spend hours *watching* television, when we could get most of the information by *listening* to it? The answer, of course, is that the visual aspect draws us into the experience so that the aural aspect makes deeper impact.

The Teacher

And as we have seen, System 5 learns best when it is drawn into an experience. It wants lots of class discussion, with an instructor who is unruffled if challenged or pressed to defend his position. System 5 challenges a teacher, not out of disrespect, but because System 5 needs give-and-take discussion. It uses that give-and-take to feel its way along and gain confidence that it truly grasps what is being said.

Where System 4 tends to see an instructor as "the authority," System 5 wants a class leader who is a mentor, a player-coach in the learning process. To learn most effectively, System 5 needs a spatial proximity between teacher and student that permits casual interaction between them. In a word, System 5 eschews the large, impersonal lecture settings that System 4 might settle for. What System 5 needs is a class small enough for everyone to see the whites of the teacher's eyes.

System 6 in the Classroom

Neither a System 4 nor a System 5 class will meet the needs of System 6. For one thing, System 6 learns best in very small groupings. Even a class that is small enough for System 5 may overwhelm System 6. Fundamentally, System 6 would limit class size to the number of people who can conveniently sit in a single circle and hear one another easily. In settings larger than that, some people will feel too intimidated to speak up. System 6, with its emphasis on consensus and congeniality, can never accept a learning experience that leaves a single person unwilling to participate. In a System 6 class, everyone is to have an unobstructed opportunity to share.

Indeed, the characteristic way that System 6 speaks of learning is in terms of sharing. System 6 classes "share" a lesson with each other. And what they share most often are personal experiences, told candidly and confessionally. System 6 groups "share" their struggles among themselves, "share" their fears and failures, "share" their dreams and disappointments. This is not some type of voyeurism on System 6's part. System 6 genuinely learns best by hearing people tell their own story.

Those stories serve two overriding purposes. First, System 6 is deeply committed to healing the wounds of the world. But since personal wounds are buried deeply, hidden from view, they must be revealed before System 6 can offer a healing hand. System 6 therefore insists on an atmosphere of trust, mutual respect, and non-judgmental listening, so that those in the group can be vulnerable and open in revealing their hurts and pains.

Second, System 6 not only wants to help those who are hurting, it also wants to understand itself. It longs to rise above the disarray of its own inner being, the distraction of its own personal needs, to position itself for selfless service of others. One of the principal ways System 6 comes to self-understanding is by following the narrative of someone else's trauma. System 6 operates on a philosophy that says, "I'll tell you my story, then you tell us yours, and everyone will gain insight from the exchange."

This all but appalls System 4, which is guarded about emotions to begin with, and would never consider sitting in a circle with other people and talking intimately about personal affairs. Not only that, the non-judgmental atmosphere that System 6 maintains, essential though it may be for openness and candor, smacks of moral indifference to System 4. As soon as System 6 starts confessing, System 4 is tempted to start preaching. To say the least, it is quite difficult to place System 4 and System 6 together in a Bible class and conduct it in a way that will satisfy both.

A Class Tailored to System 4

In the heyday of System 4 dominance, many churches standardized their Bible school curriculum so that on a given Sunday morning every adult class was teaching the same subject, using the same methodology, and following exactly the same lesson outline. Moreover, that approach was highly successful and was often accompanied by high levels of Bible class attendance. A little reflection makes it obvious why this one-size-fits-all approach becomes unworkable in a multi-system church.

To be properly systems-sensitive, we need to offer individual classes with a distinctive systems tone. The System 4 style classes would have teachers who themselves have unimpeachable credentials as Bible students. These classes would rely heavily on lecture methods and emphasize straightforward study of the biblical text. In announcing the topic of study, it would probably suffice merely to identify the section of Scripture or the doctrinal issue to be treated. Questions posed to the class would aim at determining how well participants grasp fundamental ideas in the text before them.

You could normally assume that members of these System 4 classes would arrive promptly and be ready to begin at the announced starting time. They would also be open to outside study assignments, even some occasional homework. They would not be concerned with how large the class might get, so long as the teacher was dynamic enough to maintain the greater class size.

A Class Tailored to System 5

The System 5 classes would need teachers who are held in high regard, not because they have an authoritative knowledge of the Bible, but because of their personal spiritual effectiveness. These teachers, too, must be quite competent in Scripture, for System 5 could not envision a Bible class taught by someone who does not understand the Bible thoroughly. After all, System 5 is an age of specialists. It expects authorities to know their stuff. Along the way System 5 will often cross-examine instructors, quickly exposing those whose biblical knowledge is superficial. System 5 therefore needs teachers with enough self-confidence that they do not become defensive just because students take issue with them.

Students in these System 5 classes will be more interested in learning how to get into the Bible for themselves than in what their teacher thinks about a particular passage. They want serious Bible study, to be sure, but one in which the teacher stimulates them to surface new concepts from the text. A well-planned System 5 class will use lecture sparingly. Instead, it will build on case studies, role playing, or provocative questions to trigger class discussion. It may also use "huddle groups," in which the larger class breaks into circles of four or five people to work on a project or kick around a question. In a word, a System 5 classroom is a very interactive place. The purpose of this interaction is to draw practical import from the text at hand.

Tinkering with Ideas

In the course of discussion, System 5 students are likely to toss out ideas that mark a wholesale departure from traditional ways of thinking. This reflects the "tinkering" that System 5 does when it tries to discover better ways of doing things. System 5 may or may not be serious about the ideas it tosses out. It is merely toying with these concepts, seeing where they might lead. A good teacher in a System 5 setting must be comfortable with that type of experimental thinking and not feel threatened or insecure when it occurs. The teacher's job is to return the class time and again to the language of Scripture itself, compelling students to evaluate and defend their viewpoints against what Scripture teaches.

Moreover, once the students have done that, the teacher must be prepared to step back and let them reach their own conclusions — even if those conclusions differ from cherished views of the instructor. A teacher who would be disquieted by that development, or would feel a need to impose his or her own viewpoint, will not succeed long-term with System 5 students. They want to find their own answers and to steer clear of anything that resembles spoon feeding.

Topics and Teaching Techniques

In selecting an appropriate topic for a System 5 class, we should choose one with a high potential for promoting practical insight. This does not mean that the class veers away from studying doctrine or theology. To the contrary, System 5 can manage those topics with considerable sophistication. But System 5 wants to move the discussion beyond *what* we believe to how and why it makes a difference in our day-to-day life.

As key points emerge in a System 5 class, they should be posted visually, either on a writing surface or with an overhead projector. This serves to reinforce the insights that have been uncovered. Because System 5 is a visual learner, it needs this reinforcement in a way that System 4 does not. Visuals would also be used to present charts, models, and diagrams that would further understanding. But they need to be simple and uncluttered, with concise, bulleted statements that can be quickly grasped.

With System 5 classes, unlike those in System 4, size considerations are a definite concern. System 5 classes no longer function well once the group becomes too large for effective class discussion. When the class reaches that size, we should spin off another to permit continued growth. Absent that initiative, System 5 classes tend to hit an absolute growth limit that is far below the potential System 5 "market" that is available.

To put it simply, when a class size reaches a point where members on one side of the room cannot easily understand comments by those across the way, growth will plateau. In finding new space for a System 5 class, we need to keep in mind certain basic requirements. System 5 needs a sizable classroom area (so that everyone has a clear line of sight to visuals) with considerable flexibility (so that chairs can be gathered into huddle groups, then repositioned for class discussion).

A certain amount of patience is also needed with System 5 students, for they tend to slip in just under the bell, or perhaps five to ten minutes late. How to work around this behavior, without indirectly rewarding it, is always a challenge. On the other hand, once in class, System 5 makes up for lost time. System 5 is excellent at quickly grasping new material and immediately seeing the implications it entails. They also like to flesh out ideas by talking them through. Class discussion is thus much easier to sustain with System 5 than with System 4, which tends not to comment until it is confident it has the right answer.

A Class Tailored to System 6

Once we have the proper class settings for Systems 4 and 5, what do we do for System 6? When we plan a class for them, we need a teacher who sees himself or herself as a facilitator, not an instructor. If you walk into a well-run System 6 class, it may take several minutes to figure out who the teacher is. First of all, the class is probably formed in a circle. And secondly, everyone is given unimpeded access to the discussion. To the untrained observer, the early segment of a System 6 class looks more like small talk than serious Bible study. But the small talk is System 6's way of building rapport among the members of the class, helping everyone feel a part, to sense that "I fit in here."

System 6 classes tend to personalize practical implications more than System 5 does. A typical question in a System 5 class is,

"What does this passage mean for everyday living?" The System 6 class would ask, "What does this passage mean for you?" The difference in these two questions reflects fundamental differences in the two modalities. System 5 is interested in self-expression, not self-revelation. System 5 surrounds itself with symbols of its status and achievement, but tends to keep its vulnerabilities private. After all, it has an image to maintain. System 6 is almost the exact opposite. It has no preoccupation with status or keeping up an image. Instead, it seeks out people with whom it can be open, candid, and transparent.

This is why System 5 teachers have limited effectiveness in a System 6 class. (So do System 4 teachers, for that matter.) System 5 cannot bring itself to be self-revealing, certainly not enough to put a System 6 class at ease. In a System 6 setting, if the leader is guarded about personal disclosure, the class will follow suit. As a result, a teacher who excels in a System 5 setting, may never be able to help a System 6 class coalesce.

Open-Endedness

System 6 is far less dependent on visuals than System 5. For one thing, with everyone seated in a circle, it is difficult to use a writing board or overhead projector. Besides, System 6 does not learn visually the way System 5 does. It learns from sharing life experiences with others. In a sense it learns intuitively, not didactically (as in System 4) or by experimenting with ideas (as in System 5). Therefore, a System 6 class is rich in personal story telling.

For that reason, System 6 lesson plans tend to be somewhat open-ended. Once people start talking about "what this passage means to me," we cannot anticipate where the discussion may go. For System 4 and System 5, classes need closure, a summary that says, "This is what we discovered today." That type of closure is much less important for System 6, since what students "discover" in a System 6 class is a new level of individual self-understanding. What my neighbors discover about themselves and what I discover about myself may be markedly different. There are just as many lessons learned in a System 6 class as there are people around the circle.

Matching Teaching Plans to Systems

Now that you see the stylistic difference among System 4, System 5, and System 6 classes, you can understand why each of these modalities experiences frustration in a class structured around another dominant motif. Placed in a System 4 class, System 6 will feel that everyone is interested in head knowledge, but not heart knowledge. Yet, if we put System 4 in a System 6 class, it will feel that there is too little emphasis on truth. (It does not matter what the passage means to *you*, System 4 will say. The question is, what does it *mean*.) System 4 will be equally discomfited by the level of self-disclosure that goes on around the room.

Another distinctive that sets the modalities apart in their learning style is the length of time they will pursue a topic. If we announce a year-long study of 1 Corinthians in a System 4 class, no one is likely to object. But a year-long study of *anything* overtaxes the patience of Systems 5 and 6. Even the historic pattern of examining a subject for a quarter is probably too lengthy a time frame for them. With System 5, variety is the spice of life. It needs regular doses of something new in order to remain interested. System 5 seems to respond best when a class studies a topic no more than four to eight weeks.

So how will we ever teach a lengthy book like 1 Corinthians to System 5 students? We do so by breaking the book into segments that we can treat in a two-month time frame or less. Then we give each of those segments a different title which promises helpful and practical discovery. For a class on 1 Corinthians we might announce that we are going to talk about "How To Be Different Without Being Divisive." That would take us to chapter one and the problem of schismatic factions in the church at Corinth. We could then leapfrog from that passage to others in the book, like the first paragraph of chapter three, that talk either about divisions at Corinth or offer counsel on how to work through them. Our next topic might be, "What Makes A Person Wise?" That invites a lengthy examination of the end of chapter one and the body of chapter two, where Paul distinguishes between the wisdom of the world and true wisdom.

In this manner we can substantially cover the book of 1 Corinthians, but with a notable change of pace and emphasis every few weeks. With System 6 we need not worry about changing the topic so quickly as we do with System 5. System 6 does not like to be rushed to the point that there is insufficient discussion of a subject within the group. But even System 6 can become bored by talking a subject to death, no matter how much self-revelation goes on around the circle. We might use the same approach with System 6, therefore, that we just outlined with System 5, but be willing to let things run a few weeks longer.

Childhood Education

To this point we have overlooked one vital portion of any Bible class program, i.e., the children's program. Here, too, systems-sensitivity is needed. New modalities activate in rapid-fire sequence from age five to age fifteen. By the third or fourth grade System 3 competitiveness is coming to the fore. This is a great age at which to teach the lives of biblical heroes who were movers and achievers. Students at this age need to see that being spiritual does not foreclose a person from being rugged, strong, and virile. In this regard they need to be shown a portrait of Jesus as a man of action, fearless in the face of adversaries, ready to take on wrong-doing and battle it. At this point the images of Jesus as a tender shepherd, so popular and appropriate for a younger age, should be supplemented by the portrait of a man out to change the world.

Because of the win-lose streak in System 3, competition-based learning experiences are especially effective with this age group. They are going through a critical period insofar as mastering the basic facts of the Bible is concerned. To motivate them to learn those facts, team competition in Bible-knowledge games is often highly effective. Frequent recognition and even material rewards should be offered for reaching new levels of mastery.

At about the sixth grade our fundamental approach to class should change. System 4 is turning on brightly at this point and youngsters are beginning to handle abstract concepts comfortably. If the third through fifth grade years were a time to master Bible knowledge, the sixth through eighth grade years are the time to grasp biblical doctrine. We are not talking about making education at this level some type of miniature seminary. But the curriculum should intentionally focus at helping youth at this age work through a basic understanding of such concepts as faith, grace, redemption, guilt, and Christian duty.

This is also the time to help them explore the "why" behind baptism, the Lord's Supper, worship, the church, and biblical morality. At this stage it is time to move from Jesus as a man of action to Jesus as a great teacher. In this period of early adolescence students need to explore the Sermon on the Mount and to work their way through the major parables, being led to understand the meaning behind what they read.

In today's world we have only a narrow window of opportunity in which to build this doctrinal and moral foundation. By the time students enter into high school, many of them are already adopting System 5 and System 6 thinking patterns. At that point they resist efforts at indoctrination. Indoctrination can still occur, to be sure, as it can at all stages of human development. But today's educational climate moves high schoolers quickly into values-clarification and away from values-indoctrination. By definition, values-clarification implies that a foundation of values is already in place. If it is not, values-clarification easily degenerates into moral relativism. Our object in the years immediately preceding high school, therefore, is to lay an essential foundation of morals and doctrine.

20

Words Fitly Spoken

In an unforgettable phrase, the book of Proverbs says, "A word fitly spoken is like apples of gold in pictures of silver" (Proverbs 25:11, KJV). Nowhere is that more truly the case than in systems-sensitive leadership. Word choice is critical, for each modality has specific terms that grab its attention. Through a process not fully understood, the mind becomes supersensitized to words connected to core values in our dominant modality. We actually recognize those words and process them more rapidly than we do terms identified with other conceptual systems.

Thus, if you are trying to motivate System 4, you need to pack System 4 language into what you say. If you are appealing instead to System 5, you must adjust your terms accordingly. Often we need to touch several systems with a single message. You can do that, too, by choosing your words so that you stroke each system throughout the message.

As a first step in developing your skill in this craft, you must learn to correlate the various thinking systems with words that strike a resonant chord within them. Based on what you already know about what each system esteems, you could begin building a list of those words and phrases right now. To help you get started, we will provide a listing below for Systems 2 through 7, the ones we deal with most often. Think of this list as illustrative, not exhaustive. With a little reflection you will be able to add to it. As you become an astute systems observer, you will also develop an intuitive sense of other terms and symbols that seem to spark a warm response in particular modalities.

To keep this listing reasonably brief, we have used only representative terms from an entire family of related words. For example, when you see "transcendence" under System 4, you should also think of the verb "transcend" and the adjective "transcendent." Along the way we will note certain "crossover terms." These are phrases that combine elements of two different systems. We could properly place them, therefore, in the listing for either modality.

System 2

Holy, hallowed, sacred, sanctified, saints, communion, sacrifice, atonement, purification, awe, meditation, immanent, tradition, dreams, visions, revelations, spiritual powers, Satan, demons, ghosts, haunted, seance, "the force be with you," angels, mystery, mystical, ecstasy, fantasy, the unseen world, fate, "the evil eye," defiled, unclean, cursed, walk with God, the indwelling Spirit, priest, spiritual warfare. ("Spiritual warfare" is a crossover term that also connects with System 3.)

System 3

Victory, triumph, conquer, win, master, militant, rebel, overthrow, defeat, overpower, overcome, prevail, tough, merciless, torture, strength, might, hardy, sturdy, struggle, fight, contend, do battle with, risk, humiliate, shame, retaliate, intimidate, Christian warfare, the Christian armor, redemption, ransom. (Redemption is a System 3 term because its fundamental meaning is to retrieve someone from powerless domination by another. In some New Testament passages, however, it is a crossover term. For instance, in Galatians 3:13, when Paul speaks of Christ "redeeming us from the curse of the Law," it is a crossover to System 2, which shudders at the thought of being cursed. That same passage, however, gives us another interesting crossover phrase, "the curse of the Law," which obviously connects with both System 2 and System 4.)

System 4

Truth, eternal, transcendent, absolute, heritage, doctrine, judgment, duty, responsibility, authority, law, morals, basic values, standards, policy, standard operating procedures, play it safe, stability, structure, form, integrity, ethics, fair play, self-control, self-sacrifice, discipline, rules, regulate, dependable, dedicated, trustworthy, respect, obey, submission to authority, bishop, ecclesiastical, episcopal, conversion, guilt, sin, eternal punishment, justification, innocence, forgiveness, covenant, preach, evangelize, the mind of Christ.

System 5

Achievement, recognition, reward, status, designer label, "the corporate ladder," entrepreneur, empower, options, delegate, improve, innovate, experiment, tinker, enlarge, bigger is better, fast lane, quality, excellence, professional, most advanced, successful, effective, gifted, strategy, mentor, counsel, network, self-image, high tech, cutting edge, imaginative, manage, create, how to, self-fulfillment, insightful, initiative, high energy, bottom line, created in the image of God, equip for ministry, "well-done good and faithful servant," renewed in the image of Christ, crown of life, eternal reward. ("Eternal reward" is a crossover term to System 4.)

System 6

Caring, healing, genuine, compassionate, nurturing, candor, supportive, tender, accepting, sharing, loving, consensus, shepherd, reconciliation, harmony, peace, bonding, community, environment, ecology, sensitivity, collaborate, mutual, facilitator, share your faith, bind up the wounds, the heart of Christ.

System 7

Flexible, adaptable, all-inclusive, oneness, holistic, gestalt, global, workable, process, system, the future, analyze, alternatives, quick response, visionary, all-natural, non-intrusive, big picture, interactive, informational networks, information superhighway, accepting, dynamic, all things to all people, neither male nor female.

Systems-Sensitive Promotions

When we write bulletin articles, frame announcements, or promote special events, we need to remember that each modality is supersensitive to certain key terms. If we want broad participation in a weekend seminar (System 5 likes seminars, by the way), we will spur System 4's interest by describing it as a time to deepen our understanding of Scripture. To catch the attention of System 5, we may need to mention the life skills it will enhance. System 6 will take note if we talk about the day as a transformational experience, helping us become more Christlike in our care and sensitivity.

Now, there must be integrity in a process like this. Never use language in this fashion just to gain participation. That amounts to manipulation. If the seminar will not in fact develop deeper understanding of Scripture, do not promise System 4 that it will. Find something else about the activity that will appeal to System 4 values. Or if there is nothing in the event that will truly address System 6 priorities, accept that fact and avoid System 6 language in your promotion. In addition, in a case like this do not cajole System 6 to attend. After all, you did not design the event for that modality. Otherwise the agenda would have featured distinctive System 6 elements.

On the other hand, despite the fact that the event foregoes System 6 priorities, expect some System Sixers to be present, anyway. Remember that people often have more than one dominant system. So a person who is dominant in both System 4 and System 6 may choose to attend because the program appeals to his or her System 4 strata. In systems-sensitive congregations we program toward systems. We advertise toward those same systems. Then we let the chips fall where they may.

Whenever possible, it is helpful to use phrases and promotional themes that connect with multiple systems simultaneously. One author of this book is on the staff of a church whose congregational logo has, as its most prominent element, the phrase "People Who Care." This logo contains a studied ambiguity. Seen alone, it conveys a commitment to System 6 values. But every modality *cares* about something. So we can make each modality feel treasured by tying things they value to the motto.

If we describe ourselves as "People Who Care About Morality," we create good feelings in System 4. Or we may talk about being "People Who Care About Excellence," we heartily affirm

System 5. Every time we acknowledge another set of system values by connecting them with our logo, we send a message that says, "This is your kind of place. You really belong here."

Promoting, Not Announcing

As Systems 5, 6, and 7 increase numerically in a congregation, we must become more astute at promoting events, not merely announcing them. So long as System 4 remains uncontested as the dominant modality in a church (as was the case in most congregations until recent years), promotion is not so important. Because it puts such emphasis on obedience, respect for authority, and loyalty to organizations, System 4 can be a very compliant system. Pure System 4's (that is, people whose dominant System 4 has almost no admixture of Systems 5 or 6) are "great troopers." They willingly do whatever those in charge ask them to do. In that kind of setting we can get away with being lazy promoters. Leaders merely have to announce a function, and System 4 dutifully shows up.

Systems 5, 6, and 7 look at things differently. These systems believe that they, not some authority figure, know best what they need. It does not matter to them that church leadership has decided to host a given event and has encouraged everyone to attend. These modalities will make up their own minds about taking part. As Systems 5, 6, and 7 begin to rise in a congregation, merely announcing events, programs, and activities will no longer suffice. Unless we aggressively promote them, participation will suffer. To motivate people to participate, we must help them see the specific *benefit* they will derive from their involvement. In targeting promotions at these later systems, the rule of thumb is "lead with benefits and follow with details."

Preaching to the Spectrum

In a systems-sensitive congregation, preaching makes a concerted effort to speak to all the modalities. There are vast ramifications to the concept of systems-sensitive preaching. Unfortunately, this study affords us too little space to do the subject justice. We must therefore limit ourselves to some of the more salient aspects of applying systems insights homiletically.

Preaching needs to speak to the entire person, and that means all the systems within a person. Not just the dominant systems, but all the systems. One of our principal objects in preaching is to foster spiritual health in every modality, especially those that are dominant. We want to make System 4's healthy System 4's, just as we want to make System 7's healthy System 7's. For the dominant modality to be healthy, however, the supporting systems must be healthy, too.

In chapter twenty-four, where we look at signs of unhealthy systems, we fill find that problems in one modality often root themselves in a far different modality. Rigid legalism in System 4, for example, may stem from an atrophied sense of awe and mystery in System 2. We cannot promote health in a dominant modality, therefore, unless we are also nurturing health across the entire systems mix.

Perspectives on Scripture

To connect God's Word to all the modalities, we must first learn to read the Bible from a systems perspective. In reading the Word this way, our purpose is to uncover system themes and metaphors in Scripture itself. By its own claim, Scripture thoroughly equips us for every good work. But how could it do so if it ignored systems health? Open the Bible with confidence, therefore, that you will find systems expressions throughout. Almost immediately invigorating insights and fresh meaning will begin to leap from texts that have long been familiar.

In your Bible reading make it a point to observe the language of each passage carefully, trying to identify the modality it speaks to. Look at the wording of Ephesians 6:10-17, for instance. There we see an interplay of System 2, 3, and 4. Paul tells us that we are up against spiritual forces of this world's darkness (System 2). He encourages us to don the full armor of God (System 3) and to gird our loins with truth (System 4). He also uses a number of other System 3 figures of speech that you can probably identify readily.

A few paragraphs earlier, at the end of chapter four, he touches on another combination of systems. In verse 25 we are to speak truthfully with each other, for we are members of one another. That appeals both to System 4's high sense of integrity and truth and to System 6's desire for mutuality, bonding, and candor. In the next verse we are to deal responsibly with anger, which means using System 4 principles to control System 3 impulses. Verse 27 counsels against giving Satan an opportunity. Now we are dealing with System 2 (interaction with unseen powers).

Verse 28 encourages industry and responsible labor (System 4) in order to have funds to share with those who have need (a crossover value of both System 4 and System 6). Then verse 29 insists that we should only speak words that build one another up, a quality that is attractive to System 5 with its priority on healthy self-images, along with System 6, which seeks to promote personal well-being. System 6 also responds to the "one another" thrust in verse 29. Indeed, System 6 identifies closely with all the "one another" passages in the New Testament.

Avoiding Color Blindness

At first glance this type of analysis might seem to have nothing more than curiosity value. Is there any practical benefit from being able to associate a given passage with a particular modality? As a matter of fact, there is. For one thing, it protects us from the "color blindness" of our own dominant system. When System 4 reads Scripture, it tends to see System 4. When System 6 reads Scripture, it tends to see System 6. If we are not careful, we end up drawing the themes for our preaching almost exclusively from our own dominant system. It is natural for us to do so, because those are the messages that immediately impress us as we read. But when we dig into God's word looking specifically for metaphors and dynamics from all the modalities, we are protected from becoming excessively narrow in the themes we see.

Sermon Planning

By observing the interplay of systems in a passage, moreover, we can develop more effective strategies for our lessons. If we want to teach System 4 the value of bonding with those in need (a

System 6 value), it helps to do so from a passage that has an immediate "hook" for System 4. On the other hand, to help System 6 develop deeper appreciation for diligent labor (a System 4 value), we need a text that hooks System 6. With those considerations in mind, think of how we might approach Ephesians 4:28, one of the passages we noticed above. "Let him who steals steal no longer; but rather let him labor, performing with his own hands what is good, in order that he may have something to share with him who has need."

If we were teaching this passage, especially in a limited time frame, what would we emphasize? The answer depends in large part on our target modality and the message we want to send it. If the target modality is System 4 and our goal is to reinforce System 4 values, we would stress the honest labor and doing of good that Paul implores. If our target is System 6 and we are reinforcing its values, we would anchor the main theme of our lesson in the message of sharing with the needy.

But if we were trying to teach System 4 to be non-judgmentally sensitive toward the needy (a System 6 value), our introduction should highlight the prominent System 4 themes in the passage. We could transition from the introduction by asking, "But why does the apostle urge this type of honest industry?" (System 4 loves "why" questions that are raised honestly.) We would then build the major thrust of our lesson around the need to make common cause with those in want. In other words, our strategy would be to draw System 4 into the text by emphasizing System 4 values, then point System 4 to the System 6 rationale that undergirds Paul's argument.

On the other hand, if the purpose of our lesson is to help System 6 appreciate the importance of hard work (a System 4 priority), we would reverse our strategy. We would use the System 6 part of the passage as the "hook" by constructing our introduction around the need to share with the disadvantaged. We might move into the heart of the sermon by asking, "If it is so important to share with others, how do we put ourselves in a position to do that?" And now, instead of structuring the main part of the sermon around the need to share, we would underscore the value of a responsible work ethic.

Systems and Metaphors

In a multiple-systems church, preaching must contend with the reality that theological priorities differ from one modality to the next. For example, System 4 and System 6 may both take sin seriously. But System 4 emphasizes, almost exclusively, the eternal consequence of sin. It talks about sin in terms of guilt and judicial metaphors. System 6 agrees that people can be eternally lost. But when it takes the floor to discuss sin, System 6 uses healing metaphors and stresses the alienation that sin creates in relationships. System 4 will accuse System 6 of downplaying justification, and System 6 will countercharge that System 4 treats reconciliation too lightly.

A minister can ill-afford to be caught in the middle of that crossfire. The fact is, the New Testament emphasizes justification in one passage, reconciliation in another because both are important. Nor is one *more* important than another. Yet if we are System 6 dominant and talk about the cross exclusively in System 6 language, we are certain to be criticized by System 4, and rightly so. Preaching must conscientiously pursue a balance that allows each dominant system to hear the message of Christ in its own unique language.

For instance, how would we present the doctrine of sin in a systems-sensitive manner? We would determine what systems we are addressing in our audience. Then we would portray what sin has cost us in terms of each system's existential concerns. For System 2 the cost of sin is falling under the sway of the powers of darkness. For System 3, becoming so enslaved to sin and impulse that we are no longer in control. With System 4, the loss of innocence. For System 5, the loss of the image of God. In the case of System 6, loss of intimacy with God. With System 7, the loss of harmony with the entire created order.

System Metaphors and Jesus

To extend this exercise a bit further, it might be helpful to look at the metaphors associated with the work of Christ, grouping them according to the modality they relate to. For example, the Gospels frequently depict Jesus as bringing everlasting life and releasing us from death's ultimate claim. That is clearly a message aimed directly at the concerns of System 1. So are references to Jesus as the bread of life, as well as His conversation with the woman at the well, where He held out the promise of life-giving water.

The offering of Christ as a propitiation for God's wrath draws on System 2 patterns of thought. The primary existence issue of System 2, you will recall, is whether the power behind the universe is friendly or hostile. Biblical assurances of God's friendship are meant to assuage System 2 anxieties. Purification through the blood of Christ is another System 2 metaphor. Purification is the remedy for being filthy or unclean, which arises from the System 2 sense of taboo. Whereas guilt makes a person feel bad, taboo makes us feel dirty. Passages about purification therefore have a System 2 tone.

When the New Testament talks about the redemptive ministry of Jesus, System 3 is at the forefront. This includes texts about Christ ransoming us, overthrowing Satan, and freeing us from spiritual enslavement. Redemption is deliverance from ruthless domination by a power that has taken us captive. You can see how abundantly System 3 themes figure into that motif. The promise that Jesus will receive us into glory is likewise a System 3 theme. System 3 seeks glory just as surely as System 5 pursues status. (Think of the glory given to great warriors or great athletic competitors. Or recall System 3's penchant for building magnificent capitals and monuments to celebrate its glory.) Thus, whenever Scripture assures us that glory awaits, it is stroking our System 3 emotional structure.

Imagery surrounding the cross often comes from System 4, which gives us judicial metaphors such as guilt, condemnation, forgiveness, and exoneration. System 4 also figures prominently in parables about the Day of Judgment and warnings about eternal punishment. The words of Jesus, "I am the way, the truth, and the life," are likewise System 4 themes. The sermons in the book of Acts are largely anchored in System 4, since their purpose and intent are to effect conversion among people who are System 4 thinkers.

Many of the parables borrow imagery from System 5. These

include the parable of the talents and the very similar parable of the minas, along with the parables about the hidden treasure, the pearl of great price, and the sower. Indeed, all the parables and promises about eternal reward strike a responsive chord in System 5. So do the words of Jesus when He said He came to give life in abundance. In summarizing the work of Christ, Paul was fond of picturing Jesus as a second Adam through whom we regain the image we lost in the first Adam's sin. That, too, is a System 5 theme.

The ministry of Jesus includes dozens of scenes that center on System 6 values. His miracles of healing fall into this category, along with His compassion for the multitude and His frequent defense of those who were socially marginalized. Jesus chose to describe His own ministry in System 6 terms when He pictured Himself as the Great Physician. Perhaps the most important System 6 metaphor in Scripture is the doctrine of reconciliation through Christ. System 6 themes also abound in Paul's writings, particularly his emphasis on the one body, his insistence that we maintain unity despite our differences, and his incessant concern with *koinonia*.

In chapter ten we mentioned that the flexibility the New Testament permits in local congregational governance and evangelistic methodology reflects a System 7 approach to organization. Certain statements of Jesus also embody System 7 ways of thinking. These include His use of paradoxical statements. "He who is not against you is for you," He tells His disciples on one occasion. Then a few chapters later He says, "He who is not for Me is against Me" (Luke 9:50; 11:23.) Here, as in other settings, Jesus compels His hearers to hold two seemingly contradictory ideas in tension. This type of ambiguity is characteristic of System 7 thinking.

So are the parables that compare the kingdom of God to a process of change and flux. The kingdom of heaven is like yeast planted in a loaf, like a man sowing in a field, like tares growing in the wheat. In these pictures the kingdom is not something static, but a dynamic, ever-changing reality. They anticipate the kind of world that System 7 sees, where few things are nailed down and a process of development is always unfolding.

Sermon Expectations

We must also address each modality in the way we structure sermons. When a minister enters the pulpit on Sunday morning, System 4 in the pew is asking, "What eternal verities are in this message?" System 5 wants to know, "If I apply this message this week, how will it empower me to use my giftedness more fruitfully?" System 6 will press the question, "What principles in this sermon, universally practiced, would make the world a more caring and sensitive place?" And System 7 will raise the issue, "Are these concepts flexible enough to allow us all to work together as one?"

Now, those are radically different challenges to put to a single sermon. Preaching that anchors itself in the values of only one system will leave the others untouched week after week, engendering in those neglected systems a hunger for something different. They may not be able to put their finger on why the sermon time leaves them unfulfilled, but they will nonetheless say, "It just doesn't connect with me."

The Pulpit Is Critical

No element of congregational life is more essential in promoting a systems-sensitive atmosphere than the pulpit itself. If ministers do not understand and utilize systems principles in their own ministries, the entire effort at systems-sensitive leadership will probably fail. As part of a mental or physical checklist each week, ministers need to identify the systems their sermon will affirm. While it is not necessary to stroke every system every week, neither should we be guilty of ignoring certain modalities week after week.

Sometimes, to be sure, a sermon centers on a text that only links to one or two systems. But this does not preclude us from speaking to other systems as the worship and sermon unfold. Stories and illustrations are particularly effective devices for touching systems that the sermon would otherwise neglect. Nor should we overlook the role that music can play in maintaining a systems balance during a service. If a sermon focuses primarily on concepts from one set of systems, we might use songs that honor values from other modalities.

Avoiding Haphazard Messages

The important thing is to bring intentionality to our systems messages. Whether we are systems-sensitive or not, we send messages to the various systems every time the church comes together. We have no choice, for the modalities are there, sitting in the pew, assessing everything we say and do. In a setting where systems-sensitive leadership is missing, the messages we offer each system are haphazard and catch-as-catch-can. They are sometimes also inadvertently alienating. On the other hand, by finding ways to speak a fitting word to every modality, we can help them all go away from our services and classes feeling like this truly is their kind of place.

21

The Challenge of Worship

Well-planned, systems-sensitive worship would routinely affirm every modality present. And we are speaking here not merely of each dominant modality in the pew, but of each system within every worshiper individually. Even if we are System 5 dominant, we all have System 2 needs, System 3 needs, System 4 needs, etc. Bringing the whole person into communion with God means drawing every internal system into the dialogue.

Consider the interplay of systems, for instance, when a congregation reads the Lord's Prayer in unison. This is a System 2 moment from two standpoints. First, being a prayer, it involves direct dialogue with Deity, always a System 2 event. And second, being a unison reading, it provides a setting in which System 2 is building a sense of "tribe" through shared rituals. Even the words that begin the recitation are wrought with System 2 symbolism: "Our *Father* who art in heaven, *hallowed* be Thy name."

Just because System 2 is on bold display, however, does not preclude other modalities from a vital role. The recitation is barely underway before System 4 hears words that speak to its values: "Thy kingdom come; Thy will be done." Then there follows a rapid-fire sequence of words that touch several systems:

- Give us this day our daily bread (the System 1 drive for physical survival)
- And forgive us our debts (System 4 sensitivity to guilt)
- As we forgive our debtors (System 6 compassion and commitment to the powerless, as well as System 4's sense of benevolent spirit)

And at the conclusion, the prayer builds into a crescendo of systems-stroking phrases:
- Deliver us from the evil one (System 2)
- For thine is the kingdom (System 4)
- And the power (System 3)
- And the glory (System 3)
- Forever (System 4)

One reason this prayer holds such universal appeal, no doubt, is the quiet, unobtrusive way in which it speaks at once to such a range of modalities.

Our challenge in local congregations is to structure our worship and ministry so that all the systems have an opportunity to "feel at home," as in this unison reading. Jesus once compared effective discipleship to a householder "who brings forth out of his treasure things new and old" (Matthew 13:52). Church leaders need to become like that householder. They must know how to reaffirm the "older" systems, while also providing opportunities for the newer ones to make their own special contribution. When we are successful at maintaining that balance, we tap into fresh energy and creativity throughout the church.

Neglected Systems

By being systems-sensitive in planning worship, we also lessen the risk of neglecting spiritually critical modalities. For instance, evangelical and fundamentalist churches, who are naturally inclined toward System 4 preferences, have long ignored the importance of System 2 elements in worship. When System 4 enters spiritual dialogue, it stresses structure, analysis, clear thinking, and precise understanding. System 2, by contrast, goes at truth indirectly, through symbols, story, ritual, and intuition. This approach is held suspect by many people who are System 4 dominant. Because it has fought long and hard to hold System 3 impulses and hedonism in check, System 4 has come to distrust thinking that builds on anything other than logic and reasoning.

This bias toward rationalism, imbedded deeply in System 4 culture, made its way naturally into System 4 churches. There it led to a downplaying of System 2 spiritual expression. The result has been a void in the lives of people raised in System 4 religious settings, for they have rarely felt the sense of divine encounter that is common in System 2. Nor does their worship throb with the warm kinesthetic feeling that prevails in System 2. One result of this void is that many young Christians are being drawn unknowingly into pagan notions of deity through New Age movements that know how to tap into System 2. We provide a case study of that very thing in chapter twenty-four.

Black Spiritual Expression

We might also add that black evangelical churches have done a better job than their white counterparts in keeping System 2 strains in their worship. Their highly rhythmic patterns, not just in music, but even in the characteristic pacing of much black preaching, tap into System 2. Clapping hands to the beat of the music, or swaying back and forth to musical rhythms, is also commonplace in black worship.

By marking rhythms together this way, the congregation is building a psychic sense of oneness. It is a stylized dance that unifies the tribe. It is the same type of tribe-building that a high school pep rally evokes with foot-stomping fight songs and chants. Even the hearty "amens" that punctuate black worship add to the rhythm of the service. Again, it is somewhat like a dance. The speaker leads out, affirming a System 4 truth held dear. The congregation then responds by following the movement he creates. "Preach on!" someone shouts. "We're with you," another says. "Tell it!" cries someone else. This is a System 2 tribe saying, "We are one, brothers and sisters."

This tribe-building dimension has all but disappeared from white worship except in certain fundamentalist and charismatic circles. It has remained prevalent in black churches for a number of reasons. One obvious factor is that black worshipers are only five or six generations removed from an ancestry steeped in System 2 culture. Black churches, from the very first, have drawn on that System 2 heritage, beginning with slave congregations in the Old South. By the time slave churches formed in any numbers, System 4 had dominated non-Catholic white worship for more than two centuries. White evangelical congregations were already given to the more staid, controlled environment that System 4 prefers.

This contrast between the role of System 2 in black churches as opposed to white ones contributes to the frequent failure of mergers aimed at forming an interracial congregation. No matter how well-intentioned, many of these joint efforts just do not work out. Racism is not the problem. It is the inability to transcend the significant difference between the systems mix in a typical white religious service and those where the spiritual idiom draws from the black heritage.

One footnote to this discussion. We are *not* implying that white congregations should borrow the System 2 stylistic features of black worship. There would be nothing wrong with that approach, obviously, if it proved workable. In most instances, however, bringing black styles into a white church would probably have the appearance and feel of a gimmick. What we *are* suggesting is that white churches need to find culturally appropriate ways to reintroduce System 2 motifs to their congregational life. This cannot be done hurriedly or abruptly. But it should be a principal goal of systems-sensitive leadership in churches where System 2 has been allowed to atrophy.

Putting Everybody Together

As our discussion of black and white spirituality indicates, worship is probably the most difficult place to bridge systems differences. We may be able to accommodate System 4 in one Bible class setting, System 6 in another. But they both sit together for worship. And because system preferences are so diverse when it comes to worship, it is an endless challenge to meet these competing expectations. How can we do that in a single service?

To be honest, it is not always possible. Sometimes system expectations grow so far apart that we cannot embrace all of them in the course of a single hour. And there, in the phrase "single hour," is much of our problem. Somehow, somewhere the American church became ingrained with the notion that scriptural worship ought to fit into a 60 minute time frame on Sunday morning. That persuasion took root in days when System 4 was unchallenged congregationally. When System 5 came along with its crowded schedule, it had no interest in throwing out this convention. As a result, the 60-minute worship service is an established fact of life.

If we had more time to work with, we could probably nurture several worship styles in a single assembly, rushing none of them and making the transition from one style to another somewhat smooth, not jarring. Rarely can we do that in 60 minutes. And yet, that is all the time we have. We might be able to stretch the Sunday morning hour to 70 or 75 minutes, but not much more than that. And even then we would probably leave many system needs unserved.

Multiple Service Alternatives

Therefore churches are increasingly opting for multiple worship services, not to accommodate growth, but to permit worship that is meaningful to diverse needs in the Body. This commitment to make worship enriching is a genuine sign of spiritual vitality. If a church cannot provide meaning-filled worship for its members, it needs to go out of business. A church that is not regularly drawing people nearer to God is failing in its mission. Unfortunately, the atmosphere that draws System 6 closer to God is one that may impede such closeness for System 4. And vice versa. So multiple services are the best recourse for some congregations.

Proposing a multiple service format, however, invariably triggers the objection that the church will lose its sense of oneness and fellowship if people are in different worship assemblies. But congregations by the scores have used multiple worship services without finding their fellowship impaired. When people feel that the church is effectively meeting their spiritual needs, they will be *more* loyal to the congregation, not less. And what we are trying to achieve in multiple services is precisely that — an opportunity to meet spiritual needs more effectively.

Programmed to Fail

Multiple-service formats do fail, to be sure, but not because there is an inherent flaw in the concept. In our experience they fail most often due to leadership's apologetic manner in launching them. We first saw this problem in the days when the primary purpose for dual services was to accommodate overcrowding, not diverse systems needs. All too often leaders couched the decision to add a second worship service in language like this: "As you know, we have grown so much that we can no longer seat everyone comfortably. It seems necessary, therefore, to start a second service. Beginning on such-and-such a date, we will have an early service at 8:30 a.m. and another one at 10:30. We know this will work a difficulty on some of our families, but we are sure you understand why this arrangement is necessary."

What an announcement like this does is to send a non-verbal signal that growth is bad. Growth, after all, has now thrust inconvenience on everyone by making dual services necessary. Subconsciously the congregation interprets the wording of this announcement to mean, "Our leaders are disappointed that we are growing. Maybe we should not grow any more." Time after time growth does stop right after a second service is added.

Quite often, indeed, not just a leveling off, but an actual decline sets in. The numbers continue to go down until the congregation is small enough to make a single service viable again. Critics then use the decline as evidence that the dual-service decision was ill-advised to begin with. In reality, it may have been an excellent decision. But leadership told the church indirectly, "It surely would be nice if we were small enough to fit into a single assembly." The church then complied with its leaders' apparent wishes.

Far more preferable is an approach that sincerely says, "We are thrilled that the Lord has blessed us with such growth that we are overflowing this room. That means the Lord has positioned us to grow even more and bring many more souls to Christ. For us to do that, God is obviously signaling the need for a second worship service. And so we plan to launch one on such-and-such a date. Not only do we look forward to the continued growth this will permit. But we are excited that the day may come when we need to add a third service to handle our numbers!" Now members can look at the second service, not as a congregational cross to bear, but as a stepping stone for greater service to Christ.

Avoiding Tactical Apologies

We can make a similar tactical apology, if we are not careful, in adding a new worship service to meet stylistic needs. We can leave the impression that it is genuinely unfortunate to have so much diversity in terms of worship preference that we can no longer satisfy everybody in a single setting. A second service launched with that type of non-verbal message is probably doomed from the outset. When we discussed vision-casting in chapter eighteen, we emphasized that leaders must see diversity as a *blessing*, not a curse. Not only that, leaders must be ready to celebrate diversity as a positive component of congregational life. Otherwise they can never excite a congregation about the vision of a multi-system church.

If a church opts for a second worship service aimed at specific system needs, leadership should stand before the congregation and say, "The Lord has blessed us with people who truly love to worship Him. Isn't that wonderful? And they all want the experience of worship to be as personally enriching as possible. That means, of course, that we not only want to feel enriched ourselves, but we want everyone else to be enriched, too. Our goal as a church is to provide a setting in which worship is as meaningful as possible to as many as possible. Another blessing we have is that our members find meaning in widely differing styles of worship. We are therefore planning a second service to begin on such-and-such a date. That service will have the same biblical purposes as this one. But it will use music and sermon styles that differ from what we will continue to use in this service. That will give all of us an opportunity to choose the worship setting that helps us build the closest possible relationship with our Lord."

Careful Word Choice

Notice two or three critical elements of this statement. To

assuage System 4 anxieties, it emphasizes that the second service will serve "the same biblical purposes" as the present one. Second, for those who like things the way they are, there is the assurance that their preference will still be honored in the current worship period. Third, the statement chooses language that carefully avoids anything that sounds of judgmentalism toward any set of preferences. The terms "traditional" and "contemporary" are conspicuously missing. They might convey the idea that what we are presently doing is outdated, outmoded, and old-fashioned.

Another term it avoids is "praise service." In some circles this phrase has become a synonym for worship that builds around contemporary styles of music. It easily leaves the impression, however, that historic System 4 worship is not "praise." System 4 does not take kindly to that implication, and understandably so. It immediately becomes defensive. The announcement of a second service, therefore, speaks of giving everyone an opportunity to draw near to God in a style that is not just biblical, but most natural for the individual worshiper.

Multiple Services, One Fellowship

As for the criticism that multiple services will impair unity, we need to think about what people experience in their current congregational format. If the church is larger than 150, few members, if any, know everyone. When people say that multiple services will impact fellowship adversely, they usually mean, "We will not know everybody any more." The point is, we do not know everybody now. Most of us have a few dozen acquaintances and an even smaller group of close friends whose association makes our fellowship with the church enjoyable. So long as a multiple service format does not destroy the opportunity for friendships to form and mature, we do not diminish the sense of belonging and closeness that members derive from congregational participation.

Once multiple services begin, however, it is beneficial to have periodic congregation-wide events. If space permits, perhaps a combined "fifth-Sunday worship" would be in order every time a month has five Sundays. Or maybe twice a year the church could lease a large auditorium that allows the congregation to gather as one. This serves to affirm an ongoing commitment to be one people, even if the congregation does meet in separate worship periods. And the sheer impact of bringing the whole church together at once, reminding us of our true numerical strength, has its own desirable benefit. Even if we have to crowd people and put up quite a bit of temporary seating to handle the numbers in a combined service, that in and of itself serves to heighten excitement about our growth and momentum.

Single Service Alternatives

In many instances, however, the multiple-service option is simply not workable. Some churches do not have the leadership for two distinctly different worship services on a regular Sunday morning basis. Other congregations are simply so small that dividing into two services would leave a handful of people rattling around inside a largely empty auditorium. Not only does that dampen the excitement of worshipers, it also strikes visitors as evidence of a dwindling church. Thus, if multiple services are out of the question, what can systems-sensitive leaders do?

First, they can plan every worship period to stroke each system in some way. This can be as simple as counter-balancing music and preaching. If the service is to feature a number of contemporary songs (which will appeal to Systems 5 and 6, but may not connect with System 4), we might construct the sermon to affirm System 4. For instance, we could make the sermon highly textual, even strongly expositional. Or we might use the sermon to reaffirm traditional values.

On the other hand, if the sermon is plowing new ground or exploring themes that might make System 4 uneasy, that would be a good time to sing a number of "old favorites." This sends an assuring signal to System 4 that we are not throwing out everything that has been part of our heritage. Added to that, we might choose people to lead prayers and read Scripture who have high credibility with System 4.

Keep Everyone Informed

It also helps to be "up front" about the stylistic features of a particular worship period. Explain at the outset what is planned. Then openly request forbearance (there is that word again) from those who prefer a different style. We should also reassure those who are asked to forbear that we are neither ignoring their feelings nor running roughshod over them. Instead, we are asking them to forego their preference momentarily, with assurance that we are sensitive to that preference and will honor it too in due course.

In the opening portion of a worship service, we might say something along these lines: "As you look at the program of worship today, you will notice that we are dividing the sermon into three segments. In between those segments are special songs and Scripture readings that build on the sermon themes. We will use this format for sermons through the remainder of the month. Whenever we structure a service this way, many of you tell us that you get much more out of the lesson. For others, we know, it is easier to follow a lesson when it is all in a single block, without interruption. Since we want to be sensitive to that need, too, we plan our sermons next month to be 'single block' lessons, with no songs or readings interspersed."

And while speaking of sermons, we should note how important it is to craft them carefully, with all the systems needs in mind. We developed that point at length in chapter twenty. But it seems appropriate to make it again here. Practical applications in a lesson should include something for every modality. So should illustrations and stories, which are magnificent tools for touching several systems simultaneously.

We should therefore select the illustrative aspects of a sermon carefully. If the lesson is heavy with System 4 exposition, we might draw examples from the work-a-day world of System 5 or the healing and bonding concerns of System 6. If instead the sermon is already in the narrative style that System 6 loves, we can acknowledge System 4's needs by including specific Bible references throughout it. And in the conclusion we can make it a point to underscore values dear to System 4. We may not be able to embrace all the systems with the music in a single service. But sermons, properly developed, can often do so.

Special Event Alternatives

Another alternative in a single-service format is to set a regular date (like the third Sunday of each month) on which the worship embodies a distinctively different style. Again we should bill this Sunday and promote it as a time when we are doing things in a special way out of respect for our brothers and sisters who find this approach fulfilling. If our congregation is rather traditional in tone, we might devote this special Sunday to more contemporary styles. If our church is already oriented toward System 5/System 6 preferences, our special Sunday could be given a traditional flavor, affirming the values and music that System 4 treasures.

We might also look at using services other than Sunday morning to provide an alternate worship style. That would be inconvenient, it is true. But it also faces up to an intractable reality: to meet the needs of members who want something different in worship, we must inconvenience someone. Should we inconvenience those who prefer the status quo, by asking them to accept something that goes against their taste? Or should we inconvenience those who are pressing for change by asking them to make a special time in their schedule for something different?

Faced with those alternatives, it is probably not wise to put the burden of adjustment on the ones least interested in change. Members who are themselves asking for something different may be more willing to accept inconvenience than those who like things the way they are. We might therefore plan our alternate service on an occasional Saturday evening or Sunday night. Those who enjoy that type of experience will usually rearrange their commitments in order to be present.

Whatever our scheme for special services, the key is to make them a regular part of the calendar, always done on the same cycle, so that everyone can anticipate them. Properly practiced, systems-sensitive leadership never catches people off-guard by letting them walk into a service or event unprepared for what is on tap. When something out of the ordinary is on the slate, everyone should know what is happening well in advance.

Do Them Well

Another key is to make certain that we do these "exceptional services" (i.e., those that are an exception to our normal practice) extremely well. As we plan them, we should give even more attention to detail than normal. First of all, those who do not prefer the style of this service will always be tempted to criticize. We do not want to put ammunition in their hands by conducting the service poorly. On the other hand, those who prefer worship in the style of this special service will want it done well. We cannot allow them to leave disappointed in the quality. They may conclude that we are less attentive in planning things for them than we are in planning events for others.

Worship and Vision

Whether a congregation chooses to meet diverse needs for worship in separate services or in one, the concept of systems-sensitive worship must be central to leadership's vision-casting. Since worship is the most challenging place to practice systems-sensitivity, it is vital for leaders to have clear and definite views of what they are trying to achieve in worship settings. Unless leaders can communicate their vision of worship concisely to the congregation and to newcomers, they invite endless confusion and criticism. Usually no one component of a church's life is more critical to morale and confidence than what happens in the Sunday worship. That is why it is pivotal for leaders to apply their best systems skills to that experience.

22

Maintaining Systems Alignment

In *The Fifth Discipline* Peter Senge describes what he calls an "unaligned organization."[1] Unaligned organizations have one of three qualities, he says. Either people differ in their vision of what their effort is about. Or they have incompatible mental models about how to attain that vision. Or both.

In unaligned organizations, Senge adds, leaders foster counterproductive tension when they empower groups or individuals to take initiative. The absence of shared vision means that people are not working toward the same outcomes. The absence of shared mental models invites confusion over how things should be done. In either case one element of the organization eventually ends up working at cross purposes with another. Before long, uneasiness begins to spread.

Anyone who has ever tried to resolve the tension in an unaligned organization knows firsthand that Senge is absolutely correct. But another factor not recognized by Senge also leads to unaligned organizations — the failure to maintain systems alignment. For example, in a business if management styles do not accord with the dominant modality of the work force, we have unaligned systems. Or if we use System 6 techniques to tackle a System 4 problem, we again have unaligned systems.

Churches can create unaligned systems in a variety of ways. Some of the more common mistakes occur when:

- leaders try to create an atmosphere that does not match the

---

[1] Peter Senge, *The Fifth Discipline: The Art & Practice of the Learning Organization* (New York: Doubleday, 1990), pp. 146, 231-238.

dominant modality of the congregation

- communication and motivational styles do not conform to the prevailing modalities of the members
- a ministry calls for techniques from one system, but ministry leadership functions from a different one
- verbal and non-verbal communication needlessly alienate or ignore a prominent system
- the structure and style of worship do not harmonize with the dominant modalities of the participants
- Bible classes use an approach that is incongruent with the systems-driven learning styles of the students

Anytime we have unaligned systems, conflict and strife are inevitable. Earlier chapters have already addressed several of the misalignments mentioned above. Over the next few pages we want to deal with three specific issues in maintaining systems alignment. First is the alignment of dominant modalities among key members on committees and task groups. Second is alignment of ministry leadership with the predominant system needs of the ministry itself. And third is the alignment of non-verbal messages which our ministries and organization send.

No Two Volunteers Are Alike

To illustrate how problems develop when we fail to maintain systems alignment, we might look at a simple example. Imagine that you have two volunteer positions to fill. Both involve responsibilities that will require several months to complete. For one assignment you recruit Joe, who is a System 4 thinker. Mary, a System 5 thinker, accepts the other one. First with Joe, then with Mary you clarify the outcomes that are needed. You also reach an agreement with them about the milestones their respective projects must meet.

Now, move downstream 30 days. Deciding it is time for a status report from your two volunteers, you give both Joe and Mary a call. You query them about their projects and how the effort is coming together. Anytime their answers seem vague, you press for specifics to be sure you understand everything clearly.

You then hang up, persuaded that things are going well. But a week later Mary notifies you that, regretfully, she must give up her assignment. Her rationale seems rather weak, which leads you to suspect that something else is going on. She sticks to her guns, however, and you are forced to go looking for a replacement.

So what went wrong with Mary? First, your premonitions were indeed correct. Something else *was* going on. Put yourself inside the value systems of Joe and Mary when they received your call. Joe, true to System 4 patterns, genuinely appreciated your checking up on him. Accountability is highly important to System 4. When reporting to leadership, System 4 loves to show how responsibly it is carrying out its duties. Thus, your call to Joe boosted his morale, for it gave him an opportunity to tell you about all he had accomplished. The more you asked for details, the more he felt you were genuinely interested in the good job he was doing.

Mary, on the other hand, reacted to your call from a System 5 perspective. She did not think about it the way Joe did. She did not mind you checking in to get an update from her. But when you started pressing for details, you rubbed her the wrong way. System 5 places the same high regard on competency that System 4 puts on responsibility. Your demand for specifics struck Mary as indicating a lack of confidence in her ability. No matter how polite her tone, in her mind she was thinking, "If you didn't think I was competent to do this job, why did you ask me to do it in the first place?" She hung up, not feeling affirmed, but sensing that you doubted her reliability. So she obliged by quietly resigning.

Maintaining Volunteer Morale

The contrast between Joe's response and Mary's indicates something of the management challenge in a multi-systems church. When we are coordinating the work of Systems 4, 5, 6, and 7, we cannot task them alike, supervise them alike, motivate them alike, or reward them alike. We could easily write an entire volume on the intricacies of managing volunteers from the various systems. For purposes of this introductory survey, however, we will mention only a few key points.

First, System 5 volunteers need much more recognition than we are accustomed to giving System 4 workers. Because System 4 has such a high sense of duty, we can genuinely embarrass it by giving it too much praise. System 4 feels strongly that the praise should go to God, not to the individual. Because faithfulness to responsibility is so vital for System 4, it tends to stay in the trenches and carry out its commitments, whether anyone gives it recognition or not. System 5, on the other hand, must be assured that it is giving itself to something significant. Where System 4 feels guilty if it "wastes time" by being idle, System 5 feels guilty if it "wastes time" by doing something that is unfruitful.

We must therefore pay tribute to System 5 regularly. Otherwise it begins to question how vital its work is. "If no one notices what I'm doing, they must not think it important," System 5 concludes. This is not to suggest that we give regular public acknowledgment to what System 5 is doing, but ignore System 4 and System 6. We need to maintain a steady program of recognizing what volunteers are doing, regardless of their modality. We are simply saying that it is not so consequential to forego recognition in a System 4 congregation as it is in one where System 5 is prominent.

The Motivating Word

To remain motivated, System 4 must feel that God is pleased with what it is doing. System 5 must feel that what it does is significant. And System 6 must sense that its efforts are actually changing broken lives for the better and giving people a fresh start. When we praise these systems, we need to keep in mind what is important to them. We remind System 4 of how vital their work is to the furtherance of biblical mandates. We tell System 5 how much has been achieved because of their effort. And we commend System 6 for the way its caring ministry has rekindled wearied spirits.

A sensitivity to systems differences should also govern the way we ask volunteers to take on an assignment. With System 4 we need to show that the project at hand has been thoughtfully conceived and has solid organizational structure. When we urge

System 4 volunteers to assume a new role, it is helpful to give them job descriptions that are detailed, mission statements that are specific, and expectations that are clearly spelled out. System 4 likes regular supervision and a strict system of accountability. That attention to detail leaves System 4 assured that the organization takes the work it is doing seriously. If asked to file periodic reports, written or otherwise, System 4 will normally do so ungrudgingly. And it will provide whatever information you ask for, whether it knows why you need it or not.

But where System 4 likes to have a job description and a mission statement handed to it, System 5 prefers to negotiate those items. To secure System 5's enthusiasm, we need to outline the problem to be solved, show why it is vital to solve the problem, then ask System 5 to return with recommendations for tackling the job. As with System 4, we may give System 5 a team to work with. But System 5 must know from the outset that it is free to reconfigure that team if it sees a better way of getting the job done. System 5 will want to limit its reports to what seems important. And System 5 reserves the right to determine what is important. If you ask System 5 for details it has not volunteered, you need to explain why the information is needed and what you plan to do with it.

Asking System 6 to head an effort calls for still a different technique. As a rule, the System 6 volunteer should not be given a team to head, but should be allowed to put together a team from scratch. Other modalities have difficulty on work groups and committees led by System 6. We will examine that problem momentarily. System 6 thus needs to gather a compatible group around itself, then let the group reach consensus on how to approach their task. In supervising System 6, we might as well spare ourselves the grief of expecting them to file dutiful and regular reports on their progress. System 6 is notoriously unconcerned about turning in reports on schedule, if at all. To System 6's way of thinking, time to fill out paperwork is time taken away from involvement with people. To get updates from System 6 we need to make an occasional phone call or face-to-face visit and ask for a verbal briefing.

Systems Alignment on Work Teams

In putting together teams of volunteers, we must be careful about the mixture of systems we assign to a project. As we saw in chapter seventeen, System 4 and System 6 often do not mesh well on task forces. System 4 wants to get an organization in place and have plans underway quickly. System 6 wants to forego commitment to anything until consensus starts to build. Some System 4 volunteers simply cannot serve on a committee dominated by System 6 without becoming thoroughly frustrated. After about three meetings, System 4 will be muttering under its breath, "I can't believe it. All they do is sit around and talk! How are we ever going to get anything done if we don't get organized?"

On the other hand, System 5 may fare just as poorly on a System 6 committee. For System 5 the best way to do something is to go at it as efficiently as possible. For System 6 the best way is the one that everybody feels good about. Unfortunately, those are frequently two entirely different solutions. System 5 and System 6 methodologies are also incompatible. System 5 wants to establish measurable goals. System 6 sees quantifiable goals as artificial and impersonal. System 5 wants the outcome to look professional. System 6 wants everyone to have an opportunity to contribute, even if the outcome is less than polished. Are you beginning to get the picture?

Unhappy Campers

We remember one standoff between System 5 and System 6 quite well. We dealt with it while working with the board of a wonderful church camp. This organization owned almost a hundred acres on the side of a forested mountain. The board members were primarily System 5 folk, most of them managers and high-tech engineers in fast-growing computer firms. The chairman, who himself had an engineering background, was nonetheless highly dominant in System 6. Everybody loved him, and no one would move to replace him. But his philosophy about the camp peeved the board and other supporters to no end.

For instance, when volunteers showed up on work days to prepare the camp for summer, he wanted people to work on whatever project appealed to them. If they wanted to paint, let them paint. If they wanted to build a new storage shed, let them build it. To say the least, the results were often less than professional. He would look out over the final outcome and beam. So what if trim lines were not carefully maintained in some of the painting? So what if the storage shed was not quite square? Those idiosyncrasies were evidence that everyone had pitched in and done something personally enjoyable.

The System 5 board members, on the other hand, would look at the outcome and groan, "How can we ever interest anyone in giving money to this camp when it looks like a bunch of amateurs threw it together?" Not only that, they did not want anyone to think that they, as board members, had approved anything so unprofessional. (Do you hear System 5's concern with status in their voices?) So they would come along with their own group of volunteers, trying to dress things up as much as possible.

What does this tell us, then, about structuring work groups? It tells us that within committees and ministry teams a homogeneous systems mix is often important. Groups are much more likely to maintain prolonged harmony and goodwill where their members have considerable congruency in their individual systems profiles. If Karen is System 6 dominant, we do not place her on a System 5 work group unless she has enough System 5 within her to work comfortably in System 5 settings.

Leader and Team Alignment

What is true of Karen as a work group member goes doubly for the person who chairs it. Whether we are talking of a task force, a standing committee, a study group, or a ministry, the one in charge should always represent the modality that dominates the team. Otherwise resistance will set in, covertly if not overtly. When that happens, effectiveness is inevitably undercut.

In addition, both the person chairing and the team itself need to have a dominant system that coincides with the dictates of their mission. Ministries can be thought of as systems dominant, just like people can. Evangelism, for example, is obviously a System 4 ministry, for this is the modality in which conversion occurs. Many types of benevolence, on the other hand, are System 6 ministries.

Once we identify the modality that most naturally accords with a ministry's mission, our ideal should be to staff the ministry with people who share that same dominant system. Realistically, of course, in a volunteer organization like the church, this ideal may be unreachable. So, what do we do if we cannot provide a perfect systems match between the volunteers available and the needs of the ministry?

The Chair Is Critical

Obviously we try to staff the ministry with the most interested and capable parties. But in one instance we must do our utmost to maintain systems alignment. That is in the chairmanship role. Here we need genuine alignment between the dominant system of the leader and the system requirements of the ministry itself. Appointing a System 4 deacon to head a System 6 ministry is the first step in killing it. But the same is true if we place a System 6 deacon over a System 4 ministry.

If forced to choose between a chairperson whose dominant system aligns with the ministry volunteers or one whose dominant modality aligns with the mission of the ministry, we should opt for the latter. Over a period of time leaders tend to attract others who share their systems preference. If there is systems alignment at the chairmanship level, therefore, we have an opportunity for the whole team to start evolving toward systems alignment.

Assigning Volunteers

Just as ministries struggle when we fail to keep systems aligned, volunteers do, too. They simply will not sense fulfillment in an assignment that does not tap the natural tendencies of their dominant modality. Before long they will either quit performing reliably, or they will carry out duties perfunctorily, not with excitement and enthusiasm.

People who are System 4 dominant are particularly susceptible to finding themselves in ministry assignments that do not align with their dominant modality. System 4 prides itself on being a good team player, shouldering responsibilities faithfully and carrying them out in a spirit of self-sacrifice. Thus, when asked by an authority figure to serve in a given capacity, System 4 senses a duty to comply, even if the job has little personal appeal. "If this is what leadership wants me to do, I'll be a good trooper and go along." Once it accepts an assignment, System 4 does not tend to shirk it. But in an instance like this, it may function dutifully, but without creativity and imagination.

Now, these guidelines are all rules of thumb. We have mentioned on several occasions that none of us operates exclusively from a single system viewpoint. On the inside we are multi-system creatures. A person who is System 6 dominant may turn out to be quite effective in a System 4 ministry if there is a high level of System 4 thinking in his or her makeup. Absent that quality, however, we should do our utmost to find someone else for that slot, especially if it is a critical one.

Alignment with Ministry Needs

While we are on the subject of maintaining systems alignment, it might be helpful to note certain areas of church life that play to the strengths of particular modalities. This list is by no means exhaustive. It is instead intended to illustrate the types of activities that lend themselves to the characteristics of one dominant modality or another.

We have already mentioned that evangelism is a System 4 ministry. In general, missions are, too. Medical missions, however, fall under System 6. Medical missions make the relief of suffering their primary focus, with evangelism a secondary beneficiary. System 6 is also the place we would find relief efforts, grief recovery groups, and services for shut-ins and the elderly. Singles ministries need a major System 6 component as a rule, especially those that reach out to people who are divorced and widowed. Works aimed at building bridges to other ethnic and cultural groups also benefit from a strong System 6 presence.

System 5 is the domain of counseling ministries, marriage enrichment seminars, and workshops to develop parenting skills. System 5 likewise excels at promotional activities, both internally and externally. It is especially effective when asked to rally the congregation around a campaign. With its penchant for reducing complex ideas and data to simple, uncluttered graphics, System 5 can communicate plans concisely, clearly, and compellingly. Similarly, because it feels at home with data and surveys, System 5 does a professional job on congregational research. And because it loves cutting-edge technology, System 5 is superb at developing new sound systems, video capabilities, and computer resources.

System 5 tends to fare better than other modalities at fund-raising, in no small measure because of its flair for promotions and packaging. Yet in some areas other modalities rival its fund-raising prowess. System 4, with its passion for lost souls, is good at securing support for missions. System 6 can be a capable fund-raiser if the purpose is to alleviate human misery or to provide devastation relief. In general, however, System 6 is not a good fund-raiser, nor is it a good salesman. It hesitates to be forceful in pressing for a gift or closing a sale.

When it comes to actual financial administration, System 4 is the modality of choice. Its propensity for controls, accountability, and checks and balances makes it a thorough fund custodian. System 4 is at its best in roles that demand excellent record keeping. For that reason, System 4 may be the modality of choice when we look for recording secretaries to work with a task force or ministry. System 4 is also good at tracking attendance and membership records.

As for System 7, its strengths lie in long-range planning, managing congregational diversity, and often in leadership development. It also excels at designing multi-purpose facilities and buildings that require long-term flexible usage patterns. (It may be so far-sighted, however, that others dismiss its ideas as impractical, unrealistic, or unnecessary.) System 7 needs assignments that have great variety to them, where the pattern of work changes regularly. It becomes bored with repetition. It also probably does better with assignments of rather limited duration than those that will extend through an indefinite future period. A good place for System 7 to serve is on a task force that will exist only for a finite period of time and focus narrowly on a specific problem to solve.

Ministries Needing Multiple Systems

We should also add that some ministries need core leadership from a variety of dominant modalities. Worship is one of these. The process of worship planning needs to hear input from a range of systems. Education also benefits from diverse system viewpoints, although this ministry usually functions best if its chair has the organizational concerns that are most likely found in System 4 or System 5. System 6 is often too negligent of structural details to keep a complex education ministry well-planned and administered.

Libraries typically get their start during the days of System 4 dominance in a congregation. Their initial holdings usually represent System 4 priorities. But today's library expectations probably accord more closely with a leadership profile from System 5. Be that as it may, for a library collection to be useful to the entire congregation, broad systems representation is desirable on the library committee.

Facilities management is another place where multi-system input is desirable. After all, every system in the church will be making use of the facilities. They need space, furnishings, and equipment that will serve their individual needs. If System 4 dominates facilities planning, designs will tend to be functional, a bit austere in the furnishings, and minimally attentive to aesthetics. System 4 wants to be a good steward of the Lord's money. It thus avoids anything that looks like a frill. To spend money on needless trimmings would divert funds from the essential work of missions and evangelism.

By cutting out the frills, System 4 believes, more money will be available for ministry. That often proves an unfounded assumption. If there is a strong System 5 component in a congregation, the money spent to make the plant aesthetically pleasing is rarely money diverted from ministry. This is because System 5 gives more generously to fund-raising that promises an eye-pleasing facility than it does to a campaign merely to provide functionality. Or to word it another way, cutting $50,000 out of "frills" may not result in releasing $50,000 for ministry.

Should we then avoid System 4 on facility committees? Not at all. Their frugality and dutiful sense of stewardship are critically important. But a facilities team made up solely of System 4 thinkers will not naturally provide a systems-sensitive facility. There will be little forethought, for instance, to the visual needs and flexibility requirements for System 5 programming. There will not be as much sensitivity to handicapped and disabled members as System 6 thinks appropriate. Facilities planning and maintenance should therefore be in the hands of a group that includes articulate spokesmen from all the modalities.

Aligned Non-Verbal Messages

Our discussion of facilities moves us into a final critical area of systems alignment, i.e., aligned non-verbal messages. Facilities send critically important signals that form an outsider's first deep impressions of a church. When someone drives by your facility, which system does it speak to? System 4? System 5? System 6? Which modalities will be moved to check you out because you look like their kind of place? And what about other non-verbal messages in your congregation? Do they align with who you say you are?

To walk through the issues such questions can raise, let us describe one church's experience with its facility. The congregation is located in the heart of a predominantly System 5 neighborhood, made up largely of corporate executives and professionals. But thousands of others in the neighborhood are widows and young singles. Often lonely and hungry for interpersonal bonding, they function in System 6.

Battling Architectural Limitations

Given that neighborhood makeup, it would be nice to send System 5 and System 6 signals to people who pass by. But the architecture presents a significant challenge. To put it simply, the building is a massive System 4 symbol. System 4 elevates tradition, structure, and institutionalism. And the style of this building shouts the word "INSTITUTION." Erected forty years ago, it is a magnificent structure of red-brick, fronted by towering white columns. Overhead a copper spire rises to the sky, topped with a gleaming cross. The look is so classical that advertising companies shoot commercials on the property when they need an institutional backdrop. In recent TV spots the building has been a hospital, a college dorm, and a court building.

So what message does this building send to passers-by? It obviously says that System 4 values are important to this church. And the architecture and landscaping create something of a prestigious air, which may hold an appeal to System 5. But what about all the System 6 neighbors? System 6 not only mistrusts institutionalism, it often rejects it. Since the congregation cannot tear down the building, how can it assure System 6 that this is their kind of place, too?

The approach the church took was to redesign two large signs along the front of the property. Do you recall the "People Who Care" logo we discussed in chapter twenty? It was designed specifically to target System 6 and was placed prominently on those signs. The words were painted in letters significantly larger and bolder than the name of the church itself. The first thing people see as they approach the facility is not the name of the congregation, but that message of concern and community.

The church then came back and revisited its message to System 5. The leaders wanted System 5 to walk into the building and immediately feel at home. Throughout the facility, therefore, they adopted color schemes that System 5 sees repeatedly in corporate headquarters and professional suites around town. They also added bulletin boards all along the hallways and covered them with photos and graphics, knowing how much visual displays appeal to System 5.

Being Intentional

This example shows the kind of intentionality which goes into systems-sensitive leadership, not only in the way we package our facilities, but throughout congregational life. We must continually think about what we should be doing to help each system feel valued. To that end we will obviously make carefully chosen public statements to affirm various modalities. But we must also be certain that our non-verbal messages align with those statements.

If we neglect that task, we are likely to send inadvertent messages that needlessly alienate.

For instance, we cannot imagine ourselves saying to System 5 visitors, "We are probably not your kind of people." But if we let the paint start peeling on our building, or if we permit weeds to choke the flower beds, that is precisely the message we send. We may not have intended to leave that impression. But we did so, just the same.

When we talk along these lines, some people will dismiss such "tinkering with symbolism" as somewhat irrelevant. "Does this sort of thing really work?" they ask. And the answer is, "Yes, it does work." The church we just described regularly gets calls and visits from people who decided to check them out just because of the "People Who Care" sign. One phoned to say, "I'm not a member of your church, but I drive by it each day. I'm alone, with no family nearby, and I often wonder who I would turn to if a really serious problem came up. Coming by that sign out front on my way home each day makes me feel good, because I think I could count on you in a time of need."

Other Non-Verbal Messages

While we have given major emphasis to facility messages, they are only a small portion of the total non-verbal communication in a congregation. We need to critique all of those messages continuously. To cite a single example, we can look at the relative emphasis we give to missions as opposed to being of service to the surrounding community. Mission work comes out of System 4 values. Service projects in the community typically come from System 6. As a rule congregations make much more ado about their mission work than they do their care for the elderly, the homeless, and people who are hungry. A church may frequently promote and publicize missions, but mention its work with the needy only in passing, and rarely at that. When this pattern prolongs itself, we say to System 6 in effect, "Your priorities are not important to us."

So long as System 4 is unrivaled in a congregation, it really does not matter whether we give "equal time" to both benevolence and missions. System 4 does not need to hear about helping the poor in order to feel good about the church. Not that System 4 lacks a benevolent spirit. To the contrary, it has a long and generous history of helping those in need. But benevolence is a secondary interest for System 4. Its primary concern is teaching the gospel, taking God's Word to the lost. So long as that is happening, System 4 will sense that the church is "on track," even if we never say anything publicly about our aid to the victims of misfortune. System 4, indeed, often thinks of benevolence in the community primarily as a method of opening doors for evangelism.

For System 6, on the other hand, a church has not captured the heart of Christ unless it genuinely cares for those who are deprived or suffering. System 6 needs to hear, over and over, that this church truly has a passion for the helpless. As System 6 becomes the dominant modality for significant portions of a congregation, the church must evidence its caring heart more prominently in its program of ministry, in budget considerations, and in the good works it praises and celebrates. Otherwise we say to System 6, "Maybe this is not your kind of church, after all. We never talk about the things which matter deeply to you."

In a similar vein you can also find churches that are predominantly shaped by System 6 who never talk about mission work. System 4 can feel just as alienated in that kind of congregation as System 6 does in a church that never celebrates its benevolence. Thus, leadership cannot escape the responsibility of sending messages of assurance regularly to every dominant modality in the congregation. The key, as in all things in systems-sensitive management, is to maintain balance, whatever the systems mix. Our goal is absolute accord between our verbal and non-verbal messages, along with intentionality in our messages to every system.

23

System Shifts and Transitions

While any modality can interact with another and influence it, systems-sensitive leaders pay close attention to interactions along transition zones. We introduced these zones briefly in chapter four, but have not touched on them since. They become significant, however, as we begin to analyze the dynamics of personal and organizational behavior.

Transition zones are the interval we span as we move from one dominant modality to another. Here, along these transition zones, new dominant systems gradually arise, establishing ascendancy over those that previously prevailed. Transition zones are important, for we spend so much time in them. In today's world especially, people may live most of their life in transition from system to system, not settled in a dominant modality. As individuals we cross transition zones carefully, at a measured pace, often devoting years to the process. The same is true of institutions.

Labeling the Transition Zones

In discussing systems transitions, we need to distinguish one transition zone from another. To do that, we have adopted a numbering convention that uses a simple pair of numbers. The first number represents the system where the transition begins, the second the system where it ends. Thus, Transition Zone 3-4 is the one that moves us from System 3 to System 4. Transition Zone 4-5 lies between System 4 and System 5. Over the course of a transition zone, the models from an emerging dominant system supplant the models from an earlier one. As those models vie for priority, tension between the two modalities is inevitable.

To cite a historical example, consider the challenge that faced Western Europe for more than a millennium as it transitioned from System 3 to System 4. System 3 puts its trust in power. Security, it believes, depends on being tougher than any adversary. Militarily System 3 subjugates anybody and everybody who gets in its way. Then it may humiliate them for good measure. System 4, on the other hand, trusts in principle. Truth and right, it holds, will eventually topple any tyranny, even the most ruthless. System 4 also celebrates the dignity and worth of every individual. Needless to say, this humanitarianism in System 4 creates immediate conflict with System 3's heartless exploitation of those weaker than itself.

Consequently, Transition Zone 3-4 always pits System 4 principle against System 3 power. Which one will prevail at a given moment? The answer depends on which segment of the transition zone we occupy. As we move through Transition Zone 3-4, life is always more unprincipled at the outset of the transition than toward the end. In the early stages power will commonly win out over principle. As the transition nears full bloom, principle will gain the upper hand. It is this latter portion of Transition 3-4, indeed, that tends to produce history's most respected benevolent dictators.

Transitions and the Role of Women

To offer another example, the view of women changes as societies adopt new dominant modalities. System 3 generally treats women as property. In System 3 societies, men commonly pay a bride price for their wife (or wives). Women have no public or "civic" life. They are denied entry into trades and professions. And they probably receive little, if any education.

By contrast, System 4 moves to destroy the concept of "women as property." System 4 gives them legal standing, with rights that must be protected. It also begins to provide them with something other than a domestic education. Eventually System 4 seats young girls in classrooms alongside their male peers. Despite these gains, however, women still play a subordinate role in

System 4. Except in rare instances, they never become CEOs or college presidents. On the home front System 4 treats men as "natural leaders," women as "natural followers," both endowed with those traits at birth.

That all changes when System 5 comes along. System 5 finally gives women the right to vote and to own property in their own right. In a System 5 marriage the wife is likely to be seen as an equal partner, perhaps better educated and even more successful financially than her husband. Women also sit on corporate boards, run major professional firms, and become heads of state.

When System 6 enters the picture, further change occurs. Husband-wife equality has now reached the point that many women keep their maiden name after the wedding. System 6 puts such a value on women's rights that the concept of conjugal rape enters legal theory for the first time. And outside the home, sexual harassment suits force courts and legislators to rewrite the rules for the workplace.

Relative Positions on Transition Zones

None of these changes occurs quickly, however. System 4 was dominant for centuries before the final vestiges of "women as property" disappeared. System 5 was long influential before women carved out a place for themselves in the entrepreneurial realm. And all of these system-to-system transitions have been fraught with tension over how to view women. Emerging perspectives do not win out immediately in their struggle with those they seek to replace. During those transitions we use a mixed set of models, some from the old system, others from the new one.

Thus, relative positions along a transition zone leave their mark on behavior. Not only that, a dominant system packages itself differently as we transition *into* it from the way it appears in transition *out* of it. To see this difference, we need look no further than leadership styles. Imagine two people, one of them toward the end of Transition Zone 3-4, the other in the first stages of Transition Zone 4-5. For both of them System 4 is dominant. But in one case System 3 is a junior partner to System 4; in the other case System 5 is. How will that factor affect the tone of their leadership?

The leader in transition from System 3 to System 4 will probably tend toward authoritarianism, even heavy-handedness, especially in times of conflict or crisis. This reflects the aggressiveness that comes naturally to System 3. System 3 is always eager to remind us that it is in charge. It surrounds itself with followers it can control, not followers it necessarily trusts. Thus, when a System 3/System 4 leader feels threatened, the System 3 side of the partnership presses for decisive unilateral action. Leaders like this typically make pivotal decisions without consulting anyone else. And once they announce a decision, they expect total compliance across the board.

By contrast leaders in Transition Zone 4-5 are more likely to collaborate with subordinates in times of crisis. This is the result of the growing influence of System 5. In selecting workers and associates, System 5 is more concerned with competency than control. To enlarge its own success, System 5 puts proven people in key positions, then entrusts them with broad discretion. In a word, System 5 operates at levels of trust that System 3 can never emulate. As leaders traverse Transition Zone 4-5, they become less inclined toward unilateralism when problems arise. The closer they get to System 5, the more characteristically they say, "We have talented people around here. They could probably help us with this problem. Let's see how they might deal with it."

Keep in mind that both the leadership styles we just examined were System 4 dominant. Matters of principle, duty, and stability were of uppermost concern. But the System 3/System 4 leader was tempted to be autocratic in crisis. The other leaned toward a more participative management style. To fully understand organizational and interpersonal dynamics, then, we must pay as much attention to transitional positions as dominant modalities. We can easily understand why System 4 and System 5 may disagree on priorities. But disagreement can occur just as easily when two people share the same dominant modality, but one is transitioning into it, the other out of it.

Resolving Existence Issues

Just because a person sets out on a transition does not mean that he or she will ever complete it. People and organizations commonly traverse part of a transition zone, adopt a few new models, then go no farther. Sometimes this happens because these additional models are all they need for the moment. There is no pressure to make a complete transition, for they are now able to manage the complexity of their existence. In the language of systems theory, they have reached a renewed state of homeostasis. But partial transitions may also result from unresolved existence issues. Existence issues are those fundamental concerns that define each modality. For System 3 those concerns have to do with safety and power. In System 4 they revolve around ultimate principles and absolutes. In System 5 they center on status and personal significance.

Until we resolve the existence issues of our present dominant modality — or at least feel we have done so — we will not seek out another dominant system. Thus, if people who are System 3 dominant are to adopt the principle-driven life of System 4, they must first sense that they are safe enough that they can relinquish some of their dependence on power. Similarly, System 4 will not be drawn to System 5 until the question of ultimate values is resolved.

In this regard, there is a striking similarity between the views of Clare Graves and those of Abraham Maslow. In chapter two we discussed the points of contact between the eight thinking systems and Maslow's hierarchy of needs. That hierarchy, usually represented in a pyramid structure, has been a mainstay in the fields of management and motivational theory.[1] Maslow suggested that until lower level needs are satisfied, we do not address higher level needs. Maslow held that self-actualization (which roughly equates with System 7) is the highest need that humans experience. Clare Graves disagreed with him, arguing that his own research revealed existence issues beyond self-actualization. Both men were of one

---

[1] Management texts today see Maslow's hierarchy as one of several possible ways to identify motivational categories. To see some of these side-by-side comparisons of Maslow and other motivational theorists, see J. W. McLean and William Weitzel, *Leadership: Magic, Myth, or Method* (New York: American Management Association, 1991), pp. 31-41.

mind, however, that new patterns of motivation emerge only when we have resolved the issues of the preceding state of existence.

What Resolution Entails

Since the resolution of existence issues is a necessary prerequisite to systems transitions, we should explain what "resolution" entails. It does not mean that we have in fact resolved every existence issue associated with a given modality. It only means that we have dealt with them to our own satisfaction. Some people have minimal power needs, perhaps because a fortunate upbringing presented them with few threats to their safety. They will therefore resolve System 3 issues more readily than someone who has been deeply traumatized. In another instance a person may have an intense curiosity about philosophical and theological questions, or perhaps a mind that demands rigorous answers to such questions. For this individual, resolution of System 4 issues will be a more involved process than for someone without that rigor and curiosity.

Resolution is not a "once for all" affair, however. Later developments in life may reactivate issues that we once thought settled. For instance, System 3 may resurrect its safety concerns if gangs take over my neighborhood. Or imagine another scenario, in which my doctor tells me that my disease is probably terminal. Suddenly the survival needs of System 1 take center stage.

We also revisit existence issues from earlier systems when we discover that we treated those issues too lightly. Perhaps we never fully understood the ramifications of those issues at the time we addressed them. Years later those ramifications become apparent to us, thrusting themselves upon us too forcefully to ignore. Our only recourse may be to return to those issues and think them through once more.

Downshifting between Systems

When previously-resolved existence issues resurface like this, we need emotional and psychic energy to address them. To provide that energy, we commonly "downshift" from our dominant modality. Since the later modalities are more complex than their predecessors, they require more energy to maintain. By pulling back from those more demanding systems, we free up energy and resources to fend off the new threat.

A hard-charging System 5 professional, for instance, may respond to a midlife crisis by putting career ambitions temporarily on the back burner. With the energy that move makes available, he may go back and revisit questions of values and life purposes in System 4. Or he may take up high-risk hobbies to work through power and conquest needs from System 3.

Persistent threats are another force that can lead to downshifts. If the threat proves long-lasting, we may make not one, but several downshifts as time goes by. We continue to downshift until we reach the modality from which the threat is arising. So, when a gang takes over my neighborhood, I will downshift periodically as their terror goes unchecked. To protect my home and family, I may eventually end up in System 3 myself. But since the existence issue at stake is coming out of System 3, I will not downshift further (say to System 2). To cope with threats that arise from System 3, I need as an absolute minimum the resources and tactics of System 3.

Downshifts and Stages of Dying

Terminal diseases provide a dramatic example of the downshift phenomenon. We see a close correlation between system shifts in the face of terminal illness and the stages of adjustment which Elisabeth Kubler-Ross described in her pivotal study of death and dying.[2] Suppose for a moment that I am functioning in System 7. Then I learn that I have only a couple of years to live. Since I will depend on my family to help me through the months ahead, I sense an immediate urgency to ensure that my bonds with them are strong. That means a downshift to System 6.

Next I move into what Kubler-Ross calls the "negotiating stage," where I bargain with God. "Help me beat this thing," I say,

---

[2] Elisabeth Kubler-Ross, *On Death and Dying* (New York: Macmillan, 1970).

"and I will do thus and so." This type of negotiating is typical System 5 behavior. I may also begin to think about the things I want to achieve while I still have my health. That, too, is a System 5 preoccupation.

But as time passes and the disease holds course, I sense that my negotiation has been for naught. I begin to fear that death is indeed approaching. I shift further down, into a System 4 phase, where I spend long hours rethinking, restudying, and reconfirming my faith. I also begin to think about how unfair this is (another System 4 issue), since it will leave my wife to raise our children alone.

As I focus on "how unfair this is," resentment builds until I shift down to System 3. There I may go through a period of anger at God on one hand, and a renewed determination to lick this thing on the other. Kubler-Ross noted that the period of anger with God is normally followed by a time in which we come to terms with the inevitable and make peace with the unseen forces that are dictating our destiny. That corresponds to a downshift to System 2. In this stage also, many people begin pursuing "miracle cures," even traveling abroad to try non-traditional therapies that fall more nearly in the realm of shamanism than medical science. Then, in the final stage of the disease, we live almost exclusively in System 1, struggling just to make it from one day to the next.

On a related front, the loss of a mate can cause an abrupt resurfacing of existence issues from prior modalities. Whether the loss is through divorce or death, it marks a wholesale disruption of System 2's tribal structure. Gone are the holiday, birthday, and anniversary rituals and ceremonies that provided undergirding emotional security. The need to rebuild a sense of tribe thus becomes so compelling that ambitions in other systems may be put on hold indefinitely. If a mate's loss is to protracted illness, System 4 issues may also resurface. One of these is the problem of unanswered prayer. Why did God not answer our fervent pleas for healing and recovery? If He is a kind, loving God, why did He let my mate suffer so miserably and so long? When we understand all the systems issues that surface in grief, it helps the church develop a more comprehensive response program to those who are recovering from loss.

Downshifts in Church Conflict

Most downshifts, of course, are far less extreme than those associated with adjustment to death. But even a minor downshift by a key leader can upset a delicate management balance in a church. Whenever we see a person or an organization downshift in behavior, we know that they are either sensing stress or that existence issues from previous systems have reactivated. Until that stress is relieved or the issue is resolved, the downshifted pattern of behavior is likely to continue.

Leaders who are System 4 dominant are at a particularly disadvantageous location on the systems spectrum. For them to downshift at all, their next fall back is System 3. Churches that are System 4 dominant face the same problem. One reason disagreements in System 4 churches can become so volatile is that they react to stress, as we all do, by downshifting. But that takes them immediately into System 3 and the win-lose mentality that prevails there. System 4 leaders with a high regard for principle and fair play, often resort to hardball tactics when dealing with congregational unrest. They become angry, even belligerent, hurling threats and ultimatums. That serves as evidence that a System 3 downshift has occurred.

Because this downshift to System 3 is so common in church struggles, we might say a word about how to handle it. First, it must be dealt with firmly. System 3 only respects strength. To gain its respect, therefore, you must be forceful, but not *harsh* in responding to it. You may not get System 3 to agree with you. But if you can gain its respect, you open the door for bringing about needed change. Second, while you must be firm with System 3, never do anything to make it lose face. Backed into a corner with no face-saving way to extract itself, System 3 can become ruthless. Systems-sensitive leadership never forces System 3 into that predicament.

To deal effectively with counterproductive downshifts, we must also identify the circumstances that triggered the downshift in the first place. People and organizations do not just downshift. They downshift in an effort to attain some goal. When negative behavior arises from a downshift, the goal is most commonly to fend off some threat, real or perceived. If we can help the downshifted parties attain their goal, they will no longer need to continue their undesirable behavior. If their goal is to remove a threat, we can either take measures to help them see that their fears are unfounded or else make changes that will reduce their sense of endangerment.

The Problems of Acting Out

Perhaps the most difficult downshifts to contend with are those that occur when a person becomes belligerent or autocratic as a leader in the church due to problems going on outside the congregation. Professionals who find their business getting away from them may end up "acting out" in their leadership role at church. Unable to maintain control of things in business, they shift down at church into a controlling System 3/System 4 style. The control they are trying to assert in the congregation is a means of compensating for the loss of control elsewhere. Similar types of acting out can occur when church leaders experience disruption in their family life or when they are battling psychological or physical problems they are unwilling to talk about openly.

These types of downshifts are quite difficult to manage with anything other than patience. Because the factors that are causing the shift have nothing to do with the church environment, there is little we can do in the congregational context to relieve the pressure. If this behavior becomes disruptive or demoralizing to others (as it often will), we may need to find a way to quietly slip people like this into some other set of duties where they are not so likely to cause upset.

Failing that, we must deal with the problem firmly, but again not harshly. Since this type of downshift normally results in System 3/System 4 behavior, we should use System 3/System 4 strategies to counter it. This means there should be an overt appeal to biblical principles as part of our confrontation. (This is the System 4 portion of the strategy.) In light of the biblical principles in that appeal, we should also be precise in outlining the type of conduct and behavior that simply cannot be accepted. (This draws the type of restraining line that the System 3 push for power requires.) And as always, this situation and any confrontation that may become necessary must be bathed in prayer. We may not be able to change the hearts of people, but God's Spirit can certainly do so.

Non-Adverse and Contextual Downshifts

Since we have talked about so much undesirable downshifted behavior, we should hasten to add that not all such shifts are negative. We launched our discussion of this topic by talking about several types of non-adverse downshifts. These include responses to protracted illness, or the need to revisit unresolved existence issues from earlier systems. Downshifts in and of themselves are neither good nor bad. But in interpersonal settings, one or more individuals downshifting will always have an impact on the group dynamic. Sometimes that impact is so inconsequential as to be of little concern. The illustrations above, however, include situations where the impact was indeed adverse in the life of the church.

Apart from the transitions we have noticed thus far, we should also mention perhaps the most common shifts of all. We call them "contextual shifts." They involve a behavior pattern in which we change dominant systems as we move from one context to another. We once knew a doctor who was the epitome of System 6 values if you met him in the community. But when he crossed the threshold at home, he shifted into a System 3/System 4 domineering husband. On the other hand, we have known football players in the NFL who spend hours each day in System 3, but who activate System 6 and turn into nurturing fathers when they pull into the family driveway.

In short, we cannot always observe people in one context and anticipate what their dominant system may be in another. Churches must always be sensitive to that reality. Nowhere are people more likely to make a systems shift than when they enter the doors of a church building. Unfortunately, we do not always detect that shift, for in our mind's eye a person still carries the aura that he or she has in the community.

Contextual Shifts and Leadership

We have seen policemen whose tough-minded, System

3/System 4 style on the beat gives way to a magnificent System 6/System 7 sensitivity when they take on pastoral roles in their church. At first glance we might overlook them as possible shepherding leaders, for we see them primarily as System 3/System 4 cops — Christian cops, to be sure, but cops nonetheless. On the other hand, there are people whose community aura creates an immediate impression that they would be gentle and compassionate as shepherds. If we carefully noted their system shift in church contexts, however, we might be more cautious in tapping them as leaders. We might notice that in spiritual contexts, these individuals tend to be rigid or judgmental.

To further illustrate the need to consider contextual shifts in appointing leaders, we might mention a situation we have seen on several occasions. The circumstances surround a System 5/System 6 church (i.e., one on Transition Zone 5-6). Such churches typically look for leaders who are themselves System 5 or System 6 dominant. These churches are naturally drawn, therefore, to someone who is widely respected as an innovative, imaginative professional in the community. With his high System 5 dominance, he seems an appropriate leader for the congregation. Once installed, however, he leads from a System 3/System 4 perspective. Despite his penchant for innovation in the business world, in the congregation he opposes change, thwarts efforts to delegate responsibility, and perhaps even proves rather domineering.

Everyone is taken aback. This is the last thing the congregation expected from him. Instead of an asset in leadership, he may actually turn out to be a liability, using a management style that is out of alignment with the prevailing system motif in the church. At least three factors may account for this apparent downshift. First, there may have been no downshift at all. The congregation may simply have ignored the signs that this person makes a contextual shift to System 3/System 4 in church settings. A second possibility is that he has indeed downshifted as a result of the stress he feels from this new responsibility. And third, he may be in one of those circumstances that we described earlier where he is "acting out" in his church role to compensate for difficulties in his personal or professional life.

Incremental Transitions

We should note one other aspect of system transitions before closing this discussion. At the first of the chapter we emphasized that people do not move from one dominant modality to the next in one giant leap, but in incremental steps. Once we moved to the topic of downshifting, however, our language has implied immediate and wholesale change when transitioning from one system to another. But we have done so only in the interest of simplifying our presentation. The truth is, downshifts tend to occur incrementally, just like transitions to emerging modalities do. Any problems brought on by downshifting may therefore appear gradually, over a protracted period of weeks or months. In their early stages they may go unnoticed unless the observer has well-developed system skills.

Indeed, the reason we placed this discussion of transitions and downshifts so late in the book is that we are now into regions of great subtlety. We can be effective as systems-sensitive leaders without having mastered the ability to see incremental shifts. But the more those skills become second nature to us, the more quickly we will foresee potential conflict and preempt it.

24

Sustaining Systems Health

Through most of our study we have talked about systems development as though the modalities activate sequentially from System 1 to System 8. But people sometimes bypass a modality. They may go from System 4 to System 6 with only a passing glance at System 5. Or they may go from System 3 dominance to System 5 dominance with only a cursory visit to System 4. When we skip a modality, we always run the risk of problems later in life.

As an example, a junior high teacher came up at a seminar recently, asking for advice. From the first few moments of the conversation, it was obvious that System 6 is her dominant modality. She cares deeply about her students, and she wants a classroom where everyone is affirmed and nurtured.

Unfortunately, several of her students have a different agenda. Their disruptive behavior continually plays havoc with class harmony. For a long time she believed that if she showed them enough Christian love and caring, they would come around. When their misbehavior persisted, she resorted to blaming herself for not being loving enough. Slowly but surely she finally realized that the school year was not long enough to love them out of their System 3 defiance.

A Bypassing of System 3

Now she recognized that she would never get her classroom under control unless she learned to deal with System 3 on its own terms. She had to show them she was tough and could not be intimidated. But that was impossible for her, because she had bypassed System 3 in her development. She had grown up in a strong evangelical family which believed all aggression unchristian. In the years when most pre-adolescents explore System 3 values, her parents discouraged her from highly competitive games or anything having to do with contact sports. System 3 never ignited for her. Instead, she was conditioned to feel guilt if she was ever aggressive.

Today she is trying to control teenage boys who are taller than she is. "To tell you the truth," she confided, "I am scared of them. And if I try to exert myself, I feel absolutely guilt-ridden." Having bypassed System 3, she now sorely needs it. She also needs to reorient some of System 4 values so that her System 4 judgments are healthier. She must learn that aggression is not always inappropriate. There were occasions when Jesus acted out of anger, and she can do the same thing without compromising her spirituality.

Notice, however, that we have said nothing about her dominant modality, which is System 6. Even though her problem appeared as she tried to effect System 6 values, the difficulty she faces is not primarily rooted in that system. Her dilemma is the result of an undeveloped System 3 and an improperly developed System 4. Until she builds up System 3 internally and does some retraining in System 4, her System 6 predicament will not change.

A Bypassing of System 5

To cite another example of problems resulting from a bypassed system, many of today's middle-aged adults skipped System 5 development as a result of college experiences in the late '60s and '70s. In those days System 6 bonding was the prevailing motif on campuses. Hippies, the peace movement, and budding environmentalism combined to create a System 6 atmosphere at universities everywhere. Young adults predominantly left for college with System 4 dominant and System 5 emerging, but moved quickly into System 6 in the course of their studies. If anything, they looked askance at the business community and anyone who was serious about making a dollar. In their opinion such System 5 priorities smacked of sheer materialism.

Today many of them are having to "retrofit" themselves with System 5. With retirement years not far ahead, they have few resources and no sustained career pattern which will provide for old age. Suddenly they are reading books on how to start a business, how to manage effectively, and how to make good investments. They still hold onto much of their System 6 thinking, but they supplement it now with a rising interest in System 5.

Under-Nurtured Systems

When systems activate, they ideally mature as we do. They change tone as a result of interaction with later systems. A system which fails to mature, however, can later prove as problematic as one which we bypass completely. Although any system can be undernutured, Americans are especially prone to neglect System 2 once we get out of childhood. The activist bias that pervades American values serves to discount the world of fantasy, day-dreaming, and play in System 2. As a consequence, our society conditions us to treat System 2 as relatively unimportant as we grow older. Yet problems later in life sometimes stem from an atrophied System 2.

That was the case recently with a bright young man who had interrupted a career as a high tech engineer to sort out some problems from his past. In his childhood there had been little time for System 2 development, for he grew up in an abusive, aggressive situation. System 3 turned on early for him as a necessary defense mechanism. His conversion to Christianity in late adolescence moved him rapidly out of System 3 and anchored him solidly in System 4. He was so excited about this new perspective that he gave up college to train at a Christian institute and go overseas as a missionary assistant.

Several years later he returned to the States. He resumed his college education, obtained a degree, and launched a successful professional career. By his late twenties he was well into a System 5 phase of life. He became a member of a congregation which itself has a heavy System 5 tone. His energy and drive (not to mention the overlap between the church's System 5 outlook and his own) secured him a highly visible position in the congregation.

But then problems from his troubled youth began to surface. He turned to therapy, and before long he joined a series of trauma recovery groups, one of his first major ventures into System 6. Many people he met in these recovery programs were New Age disciples. They invited him to various outings aimed at bringing about spiritual renewal. Before long he was spending weekends in the mountains, meditating in Indian sweat lodges and going through native American purification rituals. At first he thought this innocent fun. Then he began to bask in the warm, sensual afterglow of these experiences. Without understanding what was happening, he was reactivating the long-dormant System 2 of his childhood.

Slipping into Paganism

As he tapped into System 2, he thrilled to discover wonderful, empowering feelings. System 2 is an extremely kinesthetic system. He had never known anything like this before, especially in his Christian experience. Soon he began to question his own faith. Were these new sensations true spirituality, while his Christianity was a counterfeit? The longer he toyed with that question, the more he dropped his defenses. He started buying into the worldview of his New Age mentors, without even realizing he was doing so.

By the time we got to know him, he had unknowingly slipped into pagan notions of deity. Although he still considered himself a Christian and remained active as a congregational leader, he was in fact becoming pagan at heart. The strategy in counseling him was not to confront him about his System 6 New Age experiences. Instead, we opted to work with the root of the problem, which was inadequate development of Systems 2 and 4.

Retrofitting Systems 2 and 4

First, we went back into System 4 to rebuild his doctrine of God. Although he had trained at a Christian institute and had spent years on the mission field, he had mistaken an exceptional grasp of Scripture for a well-developed theology. In truth, he had never worked out a cogent, carefully-articulated doctrine of God. Once he did that, he possessed a healthy System 4 filter to use in evaluating what his new friends were saying about spirituality. Immediately he could see contradictions between what they had taught him and what biblical religion affirms.

Next, we showed him how to tap into rich System 2 moments as a Christian. Unfortunately, his religious training had been almost exclusively in System 4, with its emphasis on Bible knowledge and Christian morality. Most of his Christian life — his conversion, his time at the Christian institute, his work on the mission field, and his leadership in a System 5 dominant congregation — had been in settings that downplayed System 2.

This type of Christianity is all too common in fundamentalist and evangelical circles. It teaches little about the spiritual disciplines of constant prayer, meditation, fasting, and communion with Deity, aspects of Christian experience that all ground themselves deeply in System 2. Because he had benefited from no Christian formation of System 2, he was susceptible when he encountered System 2 in a non-Christian, spiritual context. As we worked with him, he came to the exhilirating discovery that he could keep his faith and still enjoy the embracing warmth that System 2 provides.

Once he was healthy in Systems 2 and 4, he could return to the System 6 recovery groups more profitably. They were no longer a threat to his fundamental spirituality. When non-Christian notions came up, he was able to assess them critically. He now had wholesome System 2 and System 4 models with which to evaluate new ideas. Not surprisingly, he began to make rapid progress in his recovery.

Unhealthy Systems

Throughout these examples we have seen a recurring theme. In each instance problems in a later system were traceable to underlying causes in earlier ones. We come across this pattern so often that systems sensitive managers learn to look for it instinctively.

- System 4 dominance which lacks a healthy System 2 is

commonly a combination that produces legalism. Absent the experiential and intuitive base of System 2, System 4 spirituality can degenerate into a kind of formalism and intellectualism that fosters a legalistic spirit. Having experienced little by way of awe and direct encounter with Deity, this type of spiritual expression puts its energy into understanding God's law rather than His presence.

- System 6 dominance which lacks a healthy System 4 can shed its ethical moorings. It takes little for this type of System 6 to become totally relativistic. In the non-judgmental atmosphere that System 6 encourages, standards and transcendent expectations are held in abeyance. System 6 is likely to argue for an ethic of live and let live. It therefore offers little by way of solid moral instruction.

- A transition directly from System 3 to System 5, bypassing System 4, can result in serious character flaws. In the absence of counterbalancing values from System 4, the System 3 drive to win and the System 5 drive for success merge in an unprincipled fashion. The result is often a willingness to bend the rules completely to come out on top. Many of the scandals involving inside trading on Wall Street laid bare this type of behavior.

The counseling strategy in each of these cases (or the leadership strategy in an organization) is always the same. We go back to systems that were bypassed or improperly nourished. We try to nurture those modalities to make them more supportive of overall systems strength. The same approach applies when we discover that any system with a pronounced influence is unhealthy. Some of the things that indicate an unhealthy system are:

- being closed to new ideas
- paranoia
- paralyzing anxiety
- violent outbursts
- passive-aggressive behavior
- reluctance to entertain new ideas
- rigidity
- compulsions or obsessions

These problems may appear in any system, but are usually most noticeable in dominant modalities. In addition, each modality has its own unique signs of unhealthiness. Without attempting to build an exhaustive list of these symptoms, we could mention such things as sorcery and demon worship in System 2. Cruelty and heartlessness in System 3. Excessive demands for conformity in System 4. Greed in System 5. Naive gullibility in System 6. An unwillingness to become personally involved in System 7.

You Can't Move Them Forward

As we have seen, creating health in one system is often sufficient to correct problems in others. We can abridge cruelty in System 3 by intensifying the respect for life in System 4. We can help System 7 avoid excessive detachment by enriching the impulse to bond in System 6. Neither in counseling nor in leading an organization, however, can we correct problems by trying to change someone's dominant system. As we saw early in our study, dominant systems trigger in response to our perception of complexity. And no one has the ability to change another person's internal perceptions of complexity.

Thus, you cannot solve problems in a System 4 dominant church by converting it to a System 6 church, at least not in the short run. Teaching, persistently pursued over the long haul, may eventually create conditions in which a church itself effects a move toward System 6. But there is no assurance that this transition will ever occur. And even if it does, the process will take time.

We can, however, encourage a transition that is already underway toward a new dominant system. For instance, if a transition has begun from System 5 to System 6, leaders can foster that transition by adding more and more System 6 elements to the life of the church. This should be done incrementally, over an extended period of time. The goal is to "stretch" the congregation, not disrupt it. In the early stages of a transition the current dominant system still has reservations about the one that is rising. To use traditional counseling terms, a state of approach-avoidance exists. The emerging system both attracts us and frightens us at one and the same moment.

That is why wise leaders rarely make rapid or wholesale change. Instead they pursue long-range, systematic strategies for moving in new directions. They know that homeostasis always defines some baseline to which it tries to restore balance. If sudden change seems to move the church too far from that baseline, homeostatic forces will become disruptive. They will feel a need for drastic action to bring things back to the baseline. By making incremental change, however, systems-sensitive leaders gradually adjust the baseline so that homeostasis is never seriously threatened.

In this process of "stretching" a congregation we can occasionally borrow from concepts and methodologies that are well beyond the dominant system that is emerging. All systems have the ability to acknowledge the contribution of more complex modalities. System 4 can understand System 6 thinking patterns without thinking that way itself. System 5 can borrow concepts from System 7 without adopting System 7 outlooks entirely.

It is one thing to understand a more complex modality, however, quite another to feel comfortable functioning within it. Our primary life structures will always come from our dominant system(s). Those are the ones that feel natural to us. We fall back on them instinctively when the going gets rough.

Building Strategies for Health

Because we cannot compel individuals or organizations to adopt new modalities, we must presume that their dominant system will remain in place long-term. Our goal in both teaching and counseling is to create health within the dominant modality, not move people out of it. The first issue is to determine if the current dominant modality is unhealthy. If a health problem seems to exist, we then must ascertain where the problem comes from.

- Is it the result of a previous system that was bypassed or has been left undernourished?
- Or is it the result of unresolved existence issues in the current modality?

Once we can answer those questions, we can formulate strategies with which to proceed. If the problem is in an undeveloped or undernourished previous system, that is what we try to correct. But we must do so using strategies appropriate to the dominant system. What we try to do fundamentally is to use the methodologies and priorities of the dominant system to overcome the problem we are dealing with.

To illustrate, if System 4 is lacking in System 2, we try to teach it the importance of System 2 by using what System 4 respects — straightforward Bible study. We would look at a host of passages that talk about people caught up in awe, encounters with God's holiness, and experiences with the nearness of God. We would then ask System 4 to examine its own life and determine if these qualities of spirituality have genuinely been present. (Do you see in this question an implicit appeal to follow Biblical patterns?)

Next we would let System 4 explore the issue of how one might go about creating a sense of the awe-inspiring nearness of God. Here we are letting System 4 define what it is able to accept experientially, so that we will not violate that sensitivity in the next phase of our teaching. We are also giving System 4 a chance to have the rules defined in advance, so that it knows where this experiment is taking us. (Remember that System 4 resists launching into the unknown.) Then, and only then, would we walk System 4 through some exercises that would allow it to tap into System 2 spiritual expression firsthand.

To offer another illustration, consider the problem of someone who is System 6 dominant, but does not have an adequate commitment to Biblical absolutes. To get beyond this inadequacy, we must strengthen the System 4 strata in this individual. Again, as we did in the example above, we would choose a strategy appropriate for the dominant modality, in this case System 6. We would work with this individual in a small, intimate group that develops a strong bond in the circle. We would be certain that several members of that group have very strong regard for Biblical principles.

Then we would build group discussions around the theme of "How God's Truths Have Stood By Me." We would let various ones tell stories of how they have personally triumphed over difficulty, prevailed in times of distress, or found fulfillment in the midst of loss by keeping their attention focused on a particular principle of Scripture. We might also look at how Biblical morality makes for health, happiness, and trust in human society. Our goal in these discussions is to whet System 6's appetite to learn more about what the Bible says.

Evangelistic Strategies

Systems-sensitive churches approach evangelism in a manner quite similar to what we have just described. Our objective in evangelism is to impart System 4 truth. But to get there, we must work with the dominant modality of the person we are trying to reach. We do not bring System 6 people into a System 4 lecture setting and try to teach them the gospel. Conversely, we do not put System 4, with its emotional reserve, in an interpersonally intense System 6 setting and try to lead them to Christ.

Instead, we make our evangelistic approach multi-faceted, just like our educational program. If we are working with System 5 people, we use System 5 strategies to connect with them, build relationships with them, and then through those relationships take them back into System 4.

We do the same thing with all the other systems. If we are working with tough inner-city neighborhoods where System 3 prevails, we do evangelism through people who are strong, virile, and fearless. They talk freely about their faith in Christ and how it has given them the courage to face any danger, to stand up against all wrongdoing. Those qualities, both in the messenger and the message, connect with System 3 and allow us to build a relationship through which evangelism can move forward.

Missions Strategies

In missions, too, it is important to use appropriate systems strategies. Drawing on the experience of Don Browning, who co-authored this book, we could point to the challenge of taking the gospel to System 2 cultures. He has had the experience of preaching in those settings repeatedly in the Third World. What he finds is that the gospel is less effective in such circumstances if preached in the System 4 categories we are most accustomed to in the U.S.

— guilt, judgment, justification, forgiveness, etc. What connects with those audiences is System 2 imagery from Scripture. Jesus as the one who protects us from God's wrath. The cross as purification from spiritual uncleanness. The power of a holy God whose holy things must be respected.

Over the past thirty years countless works have been written on cross-cultural evangelism.[1] Systems-thinking gives us another aspect of communication to think about when we must transcend cultural lines with the gospel. Because Scripture describes the work of Christ in metaphors that are drawn from every modality (as we saw in chapter twenty), we can tell the wonderful story of Jesus in any system context, using a frame of reference that strikes a resonant chord in that system, without ever violating God's truth.

A Final Word

Indeed, what we have just said sums up the essence of systems-sensitive leadership — connecting Christ to individual lives within the context of dominant systems, yet staying true to God's Word and His will in the process. We firmly believe that the Word of God, properly taught in systems-sensitive settings, can create health in all dominant systems.

- It can help System 2 experience the realm of spiritual realities without lapsing into superstition.
- It can help System 3 be tough without being heartless.
- It can help System 4 delve deeply into truth without becoming a legalist.
- It can help System 5 pursue the image of Christ without succumbing to the image-consciousness of its world.
- It can help System 6 be caring and compassionate without developing a sense of spiritual superiority.
- It can help System 7 work on "big picture" issues without losing a heart for hands-on involvement in the small, seem-

---

[1] Understandably, most of these works have come out of the field of foreign missions. One of the most thoroughgoing is David J. Hesselgrave, *Communicating Christ Cross-Culturally* (Grand Rapids: Zondervan, 1978).

ingly insignificant acts of ministry that may not turn the world around, but can turn a life toward God.

Systems-sensitive leadership is not a panacea. We would never suggest that it is. It does not eliminate all conflict and friction from the church. It does not preclude clashes of personality or the pettiness of childish attitudes. But it does provide exceptional tools for creating health, harmony, and happiness within a Christian community.

As we have put the concepts of this book into practice ourselves, we have immediately seen their positive impact. We hope you will have an equally exhilirating experience as systems-sensitivity becomes part of your own ministry. We recognize that our examples and guidance have been far from exhaustive. There is much more to be said on every subject we broached. But our purpose was to offer an introductory application of systems-thinking, not a full-fledged exposition. Other books may follow later that will elaborate extensively on topics that demand greater attention than we have been able to give them here.

Our prayers go with you as you work toward harmony and oneness in your congregation. The challenges to unity have never been more demanding, the importance of unity never more urgent. May God prosper you richly as you work to create a peaceable kingdom, and may His Spirit always sustain you as you lead His people and His church.

About the Authors

Since 1986 **Michael Armour** has been the pulpit minister for the Skillman Church of Christ in Dallas, Texas. Prior to that he was dean of student affairs at Pepperdine University and president of Columbia Christian College. His 30 years of ministry have stressed counseling and leadership development. As a captain in the naval reserve, he specializes in naval intelligence and computer technology. He holds graduate degrees in religion and history, including a Ph.D. from UCLA.

Don Browning is a native of West Texas and received his formal education at Abilene Christian University. He was awarded the B.S. Degree in Bible in 1959, and did additional graduate work in Bible at A.C.U. and Wesley Theological Seminary in Washington, DC. He has served the Singing Oaks Church of Christ in Denton, Texas for the last 13 years.

Don is actively involved in mission work in India. During the past 20 years, he has traveled to India 18 times, which provided him a laboratory for observing multiple thinking systems functioning simultaneously in society. Opportunity for testing the Gravesian model in ministry led to this writing project. He has also spoken on several Christian College Lectureships and teaches seminars and workshops.

Don and his wife Charlene have two daughters and a grandson.